This
LAND

57
WAYS

*To Protect Your Home Environment
(and Yourself)*

Rick Weinzierl
Patrick Weicherding
Brenda Cude
David Williams
Doug Peterson

*Produced by Information Services
College of Agricultural, Consumer and Environmental Sciences
Cooperative Extension Service, University of Illinois at Urbana-Champaign*

North Central Regional Extension Publication 583

North Central Regional Extension resources are subject to peer review and prepared as a part of extension activities of the thirteen land-grant universities in twelve north-central states, in cooperation with the Cooperative State Research, Education, and Extension Service (CSREES), U.S. Department of Agriculture, Washington, D.C. The following universities have approved this resource for regional status: University of Illinois at Urbana-Champaign, Kansas State University, Michigan State University, University of Minnesota, University of Missouri, Ohio State University, University of Wisconsin.

For copies of this and other North Central Regional Extension resources, contact the distribution office listed below for your state. If your university is not listed, contact the publishing university (marked by an asterisk).

*University of Illinois	Michigan State University	University of Wisconsin
Ag. Publication Office	Bulletin Office	Cooperative Extension Publications
67 Mumford Hall	10B Ag. Hall	Rm. 245
Urbana, IL 61801	East Lansing, MI 48824-1039	30 N. Murray St.
(217)333-2007	(517)355-0240	Madison, WI 53715-2609
		(608)262-3346

*Publishing university

Programs and activities of the Cooperative Extension Service are available to all potential clientele without regard to race, color, national origin, age, sex, religion, or disability.

In cooperation with the NCR Educational Materials Project.

Issued in furtherance of Cooperative Extension work, Acts of Congress of May 8 and June 30, 1914, in cooperation with the U.S. Department of Agriculture and Cooperative Extension Services of Illinois, Indiana, Iowa, Kansas, Michigan, Minnesota, Missouri, Nebraska, North Dakota, Ohio, South Dakota, and Wisconsin. Dennis R. Campion, Interim Director, Cooperative Extension Service, University of Illinois at Urbana-Champaign, Urbana, Illinois 61801.

January 1996

Printed and distributed in cooperation with Cooperative State Research, Education, and Extension Service (CSREES), U.S. Department of Agriculture, Washington, D.C., and Alabama, California, Colorado, Delaware, Idaho, Mississippi, Nevada, Oklahoma, Oregon, South Carolina, Tennessee, and Vermont.

The information provided in this publication is for educational purposes only. References to commercial products or trade names do not constitute an endorsement by the University of Illinois and do not imply discrimination against other similar products.

ISBN 1-883097-10-X

This publication is printed with soybean ink on recycled paper.

This Land series coordinators: Michael Hirschi, UI Department of Agricultural Engineering; Doug Peterson and Gary Beaumont, Information Services

Technical coordinators: Brenda Cude, UI Department of Agricultural and Consumer Economics; Rick Weinzierl, Patrick Weicherding, and David Williams, UI Department of Natural Resources and Environmental Sciences

Principal writer: Doug Peterson

Assistant writers: Marilyn Upah-Bant, Gary Beaumont, Kay Bock, Dave Petrina

Editor: Nancy Nichols

Designers: Marisa R. Meador, Susan Skoog, Myung Lee

Technical illustrator: M.R. Greenberg

Proofreaders: Bernard Cesarone, Cheryl Frank, Phyllis Picklesimer

Support staff: Irene Miles, Joli Shaw, Oneda VanDyke

Photo credits:

Page 2 Jonathan Fickies, *Binghampton Alumni Journal,* New York
12 Courtesy of Ryobi America Corporation, Anderson, South Carolina
13 Brian Stauffer, Information Services, with cooperation from Lowe's, Champaign, Illinois
51 Courtesy of the *Columbus Dispatch,* Ohio
61 Courtesy of Gardens Alive!, Lawrenceberg, Indiana
90 Brian Stauffer
98 Brian Stauffer
99 Brian Stauffer
100 Brian Stauffer
126 Courtesy of the Waste Watch Center, Andover, Massachusetts
149 Brian Stauffer
182 Courtesy of the Des Moines Water Works, Iowa
191 David Riecks, Information Services
217 Brian Stauffer
231 Tami S. Dalto, courtesy of Vernon Communications, Inc., Bellevue, Washington
255 Courtesy of Shelter Supply, Minneapolis, Minnesota
274 Brian Stauffer
275 Brian Stauffer
276 Brian Stauffer

Technical reviewers:

The following people reviewed drafts of the text in its entirety.
University of Illinois at Urbana-Champaign
Brenda Cude, Extension specialist, consumer economics
Nicholas Smith-Sebasto, Extension specialist, environmental education
Patrick Weicherding, Extension specialist, urban forestry
Rick Weinzierl, Extension specialist, entomology
David Williams, Extension specialist, horticulture

The following people reviewed portions of the text.

University of Illinois at Urbana-Champaign
Robert Aherin, Extension specialist, safety
Mel Bromberg, Extension specialist, drinking water and health
Kathleen Brown, Extension educator, community leadership and volunteerism
Poo Chow, professor of wood science
Darin Eastburn, Extension specialist, plant pathology
Rhonda Ferree, Extension specialist, horticulture
Mary Ann Fugate, Extension educator, consumer and family economics
Floyd Giles, Extension specialist, horticulture
Dawn Hentges, Extension specialist, food safety
Michael Hirschi, Extension specialist, soil and water
John Lloyd, Extension specialist, entomology
Mike McCulley, School of Architecture—Building Research Council
Sandra Mason, Extension educator, horticulture
Philip Nixon, Extension specialist, entomology
Nancy Pataky, Extension specialist, plant pathology
Stephen Ries, Extension specialist, plant pathology
James Schmidt, Extension specialist, horticulture
Henry Spies, School of Architecture—Building Research Council
Thomas Voigt, Extension specialist, horticulture
Christine Wagner-Hulme, graduate assistant, entomology
Robert Wolf, Extension specialist, agricultural engineering

Illinois Department of Conservation
Division of Natural Heritage

Illinois Department of Energy and Natural Resources
Charles J. Russell, manager, consumer assistance section

Illinois Department of Nuclear Safety
Marjorie Walle, manager, and Cindy Ladage, policy analyst, Radon Program, Environmental Monitoring Section

Illinois Department of Public Health
Tom Long, toxicologist
Clint Mudgett, chief, Division of Environmental Health

Illinois Environmental Protection Agency
Les Morrow, environmental toxicologist

Illinois Hazardous Waste Research and Information Center
Daniel Kraybill, technical assistance engineer

Illinois State Water Survey
Brian W. Kaiser, chemist
Loretta M. Skowron, director, Office of Analytical and Water Treatment Services

Community Recycling Center, Champaign, Illinois
Matthew Snyder, executive director

University of Nebraska-Lincoln
Shirley Niemeyer, Extension specialist, home environment

Purdue University
Robert M. Corrigan, Extension specialist, animal damage control

Rutgers University
Joseph Ponessa, Extension specialist, housing and energy

Solid Waste Agency of Northern Cook County
George W. Brabec, director of recycling and waste reduction

Waste Watch Center, Andover, Massachusetts
Dana Duxbury, president

Contents

Making the environment a household word 1

Caring for the home landscape
   1. Aim for a healthy lawn rather than a perfect lawn 4
      Coming to terms with pests (and IPM) 11
   2. Cut the grass without gas 12
   3. Prevent insect, disease, and weed problems in the garden 15
   4. Identify fruits and vegetables that need fewer pesticides 19
   5. Select native trees and shrubs that will thrive in your yard 21
   6. Keep your trees healthy and safe 27
      Is treated wood safe? 32
   7. Plant "wild" in your backyard 33
   8. Attract birds and butterflies to your backyard 41
      What about the wildlife you don't want? 47
      Life down on the butterfly farm 51

Using alternatives to traditional pesticides and fertilizers—where practical
   9. Scout for pests on the landscape 54
  10. Consider microbial insecticides 56
      Organic pesticides: Patience pays off 61
  11. Consider botanical insecticides and insecticidal soaps 62
      Don't play chemist 66
  12. Conserve beneficial insects 67
      Do companion plants deter pests? 72
  13. Monitor and control insects with traps 73
      Don't get zapped by unreliable pest-control devices 76
  14. Use pest barriers in your garden 77
  15. Solarize soil 79
  16. Consider the potential of natural fertilizers 81

Using chemicals on the landscape—if necessary

17. Know what to ask a lawn-care company and arborist 86
    Lawn-care industry moves toward less toxic products 90
18. Understand pesticide toxicity 92
    A pesticide primer 95
19. Read the pesticide label 97
20. Apply pesticides accurately and safely 101
21. Apply fertilizers accurately and safely 108
    Calibrating your application equipment 112

Storing and disposing of hazardous chemicals

22. Store hazardous chemicals safely 116
23. Dispose of pesticides safely 121
24. Dispose of auto products safely 123
    Keeping an eye on hazardous waste 126
25. Dispose of paints and solvents safely 128
    Solvent-free paints stir up changes 130
26. Dispose of adhesives, aerosols, household cleaners, and other hazardous waste safely 131
    Sorting out the battery dilemma 134

Managing yard waste, food waste, and other household waste

27. Reduce household waste 138
28. Recycle paper, glass, and aluminum (and a few other household wastes) 141
    Keeping contaminants out of the recycling stream 147
    Small recyclers work together to make trash pay 149
29. Reuse and respond 151
    Environmental trade-offs 153
30. Recycle grass clippings 154
31. Create a compost pile 156
    Types of composting bins 160
32. Use yard waste as mulch 162
    Other organic mulches 164
33. Dispose of tree residues 166

Protecting your drinking water

34. Know when and how to test your water 170
    How to collect a water sample 173
35. Test your water for coliform bacteria and nitrate 174
    Blue-baby syndrome 176
36. Test your water for pesticides—if necessary 178
    *The ins and outs of drinking water: What goes into it,
    what's taken out* 182
37. Know the signs of contaminants in drinking water 186
38. Select effective water-treatment methods 188
    Water-treatment scams 194
39. Prevent contamination from septic systems 196
40. Know the pros and cons of bottled water 198
    *Sparkling, natural, mineral, spring, artesian...Are you
    confused?* 201

Protecting the indoor environment

41. Prevent contamination by lead 204
42. Test your home for radon 208
43. Reduce radon levels 212
44. Control dust, dust mites, and other allergens 216
    Make sure humidifiers do more good than harm 220
45. Reduce pollution from combustion equipment 221
    Carbon monoxide Q & A 224
    Double messages and back drafts 226
46. Prevent contamination by formaldehyde and asbestos 227
    Seattle residents put their homes to the test 231
47. Recognize that it's not always necessary to control
    indoor insects 234
48. Select the right household pesticide 240
49. Control household pests with less toxic alternatives 242
50. Select household cleaners wisely 246

Conserving energy and water

51. Design new homes for energy efficiency 250
52. Save energy in heating and cooling 252
    Cutting the cost of keeping out the cold 255
53. Insulate and seal your home 258
54. Choose energy-efficient appliances 266
55. Save money and energy with compact fluorescent light bulbs . . . 272
56. Use plants and other landscape tools to conserve energy 279
57. Conserve water in and around the home 283

For more information 289

Home environment goals 295

Index 303

This LAND

Making the environment a household word

For the Burger family of Whitney Point, New York, taking out the garbage has become something of a media event. After all, it happens only once every five years.

Chris and Cindy Burger have a fairly straightforward system for cutting down on garbage. They look for products that use little or no packaging, and they separate glass, plastic, metal, and paper for recycling. Nonrecyclable paper is chopped up and dumped in their compost pile, which is also the final resting place for food waste.

But what about those troublesome items, such as styrofoam and certain plastics? The Burgers have a friend who lives near a factory that takes styrofoam, and Chris's parents live near a firm that takes all plastics. So whenever they visit, they take a load with them.

Cutting household waste to one can of trash every five years may sound like an impossible goal to most people, but it demonstrates just what can be done—especially if you take it one step at a time, as the Burgers have been doing since 1970.

The rewards will be there, because what you do to improve your home environment can make an immediate and sometimes dramatic difference. For instance, the U.S. Environmental Protection Agency has identified indoor air pollution as a high environmental risk. But you can reduce the risk by lowering the levels of lead, dust, radon, and other contaminants in your home.

57 Ways to Protect Your Home Environment (and Yourself) will help you to make these improvements, as well as many others—such as cutting back on pesticides in the yard and safely disposing of the hazardous waste collecting in your garage. As a bonus to improving health and safety, many of the practices described in this book will not cost you anything. Some of them may even save you money, especially those in the energy conservation section.

Case in point: Switching from traditional light bulbs to compact fluorescent bulbs can save substantially in energy costs. Over the life of one compact fluorescent bulb, you can save $54 for every light fixture in the house. (See page 272 for more details.)

In addition to the personal advantages, these practices offer wider benefits—less stress on the waste-handling system and local water-treatment plant, less risk of groundwater and surface-water contamination, and a healthier environment for birds, beneficial insects, and other wildlife.

You probably will not need to do everything described in this publication, and you may not *wish* to do everything. But do what you can. Scan the chapters and decide which sections you wish to focus on first. Then outline a plan of action and follow up on it. The checklist on page 297 will help you set up such a plan.

As some of these changes become part of your life, they may even become habit-forming. Chris Burger likens it to driving a stick shift.

"People who drive an automatic have the idea that you have to think about shifting gears all the time," he says. "But once you get into the habit, it becomes second nature. It's not as daunting as you might think."

Thanks to their recycling efforts, the Burger family fills a trash can only once every five years. From left: Jennifer, Chris, Cindy, and Debbie.

Caring for the home landscape

"I care for my lawn about as well as Godzilla cared for Tokyo."
—Dave Barry, syndicated columnist

What makes an ideal lawn or garden will vary from person to person because some are looking for a conventional landscape, whereas others have more adventurous goals. In this section, you will find some useful hints for both the conventional and the nonconventional yard, as well as ideas on how to create a healthy landscape with a minimum of synthetic chemicals.

Chances are, there are even ideas here for those who subscribe to the Dave Barry School of Yard Care.

1 Aim for a healthy lawn rather than a perfect lawn

Develop reasonable expectations

A typical lawn of 5,000 square feet contains approximately four million turfgrass plants, making it difficult to sustain a high-quality lawn for long periods. You will save yourself a lot of work if you simply concentrate on creating a healthy lawn, rather than the manicured, Wrigley Field look. You may also be able to cut down on pesticide and fertilizer use because they are often overused in the quest for a perfect lawn.

One of the best ways to create a healthy lawn is to practice *integrated pest management,* better known as IPM. An IPM program aims not to eliminate all pests but to maintain pests at tolerable levels. Pesticides are often part of an IPM program, but they are not used unless other control methods are unable to produce the desired results.

An IPM program focuses on several key points to develop a healthy lawn: site evaluation, soil preparation, turfgrass selection, fertilization, thatch removal, mowing, watering, physical or mechanical removal of pests, prevention and sanitation, and pesticides and their alternatives.

Site evaluation

Heavily shaded areas do not offer good growing conditions for turf, and pest problems can result. Severely sloped areas, meanwhile, are often difficult to establish from seed, and once they become established, severe slopes can be difficult and dangerous to mow. Therefore, ground covers other than turfgrass are better suited to these areas.

In heavily shaded areas, for instance, consider such alternatives as pachysandra (Japanese spurge), English ivy, periwinkle, lily of the valley, and hosta. Be sure to select varieties that adapt to your area. For severely sloped areas, options include day lilies, birdsfoot trefoil, or low-maintenance native grasses.

Soil preparation

Many turf problems can be traced to soils that are excessively wet, dry, acidic, or alkaline, and to soils that are infertile, compacted, or full of debris. The best time to modify your soil is when you first establish the lawn. Here are some ideas:

- Have your soil tested, then add fertilizer and lime or sulfur if indicated by test results and recommendations.
- Remove all construction debris and rocks from the soil.

- If you have added fertilizer, lime, or sulfur, till it into the soil to a depth of 6 inches. Loosen soil that has been severely compacted by construction equipment.
- Direct drainage away from the house, and fill in depressions that will collect water.
- If the existing soil contains a lot of sand or clay, mix 1 to 2 inches of a good organic matter source (such as peat or humus) into the top 4 to 6 inches of soil.

Turfgrass selection

Carefully select turfgrasses based on the use, growth environment (soil type, sunlight availability, moisture), desired appearance, pest resistance, and management level required. Different grass species offer different qualities. Even within a species, there are significant variations among the different "cultivars" (grass varieties).

If you select a low- to moderate-maintenance grass variety, you will have a more durable lawn that requires less frequent mowing, watering, and fertilization. In addition, a low-maintenance lawn is often more resistant to pest problems and, therefore, less dependent on pesticides. A low-maintenance lawn, such as tall fescue, may not be as attractive as a high-maintenance turfgrass, such as Kentucky bluegrass, but you may not notice the difference unless you see the lawns side by side.

To get an idea of some qualities of popular grass species, refer to the accompanying table, "Common grasses." Also, to find out what varieties have special disease resistance, contact your local Cooperative Extension Service office.

Fertilization

A properly fertilized lawn will maintain good color and density, and will grow rapidly enough to discourage weeds and other pests. Conversely, inadequate fertilization can result in a sparse lawn, which can be invaded more easily by weeds.

For low- to moderate-maintenance lawns, you can get by with two nitrogen applications per year, whereas higher quality lawns require three or four applications. For homeowners in the Midwest, the table on page 6 shows the best times to apply nitrogen on cool-season turfgrass, depending on how many applications you intend to make.

Leaving grass clippings on the ground to decompose is one good organic source of nutrients, especially nitrogen. In fact, some mulching-mower users have found that they can cut yearly nitrogen applications by one-fourth when grass clippings are left on the lawn.

Common grasses

Note: For information on the resistance of particular varieties to disease and insects, contact your nearest Cooperative Extension Service office.

Kentucky bluegrass
- Medium tolerance to wear and readily recovers from damage.
- Full sun to light shade.

Perennial ryegrass
- Good tolerance to wear, but recovers poorly from damage.
- Full sun. Avoid temperature extremes. Rarely used alone in the Midwest.

Tall fescue
- Good wear tolerance, but recovers poorly when damaged.
- Full sun to light shade. Does well in heat and drought. Grows best in the southern two-thirds of the Midwest.

Fine-leaf fescues
- Moderate tolerance to wear and poor to moderate recovery following damage.
- Usually does well in dry locations with light to moderate shade.

Zoysiagrass
- Good wear tolerance and a moderate to good recovery rate following damage.
- Full sun to light shade. Tolerates heat and drought well. Grows best in the southern part of the Midwest.

Nitrogen applications on cool-season turfgrass

Number of nitrogen applications per year	Time of application			
	First	Second	Third	Fourth
One	Early to mid-September			
Two	Late April to early May	Early to mid-September		
Three	Late April to early May	Early to mid-September	November	
Four (only with summer irrigation)	Late April to early May	Mid- to late June	Early to mid-September	November

> ✓ **CHECK IT OUT**
> For more information on using grass clippings as mulch: page 162.
> For information on alternatives to traditional, synthetic fertilizers: page 81.

Thatch prevention and removal

Thatch is the intermingled layers of living and dead stems, leaves, and roots that exist between the soil and green vegetation. Excessive thatch—½ inch or more—can provide a protective home for some insect pests and disease-causing fungi. It can also prevent certain pesticides from penetrating to the soil level, making them ineffective. Thatch can even prevent air and water from reaching the soil and root zone, reducing the turf's tolerance to drought and temperature extremes.

You can prevent or remove thatch with these practices:

Proper watering and fertilization. The best way to prevent excessive thatch buildup is to avoid overwatering and overfertilization.

Core aerification. With this process, you insert hollow tines into the soil and remove soil plugs throughout the lawn. Core aerification does not

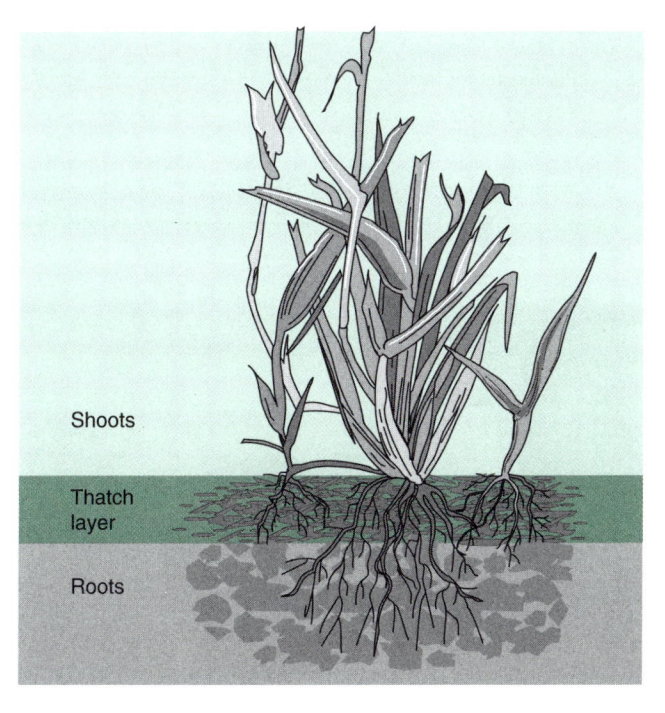

Thatch is an intermingled layer of living and dead plant material that collects on the soil surface of your lawn.

remove much thatch, but it helps to prevent thatch buildup and to improve the soil's condition.

Mechanical dethatching, power raking, and vertical mowing. These mechanical methods use a series of spinning vertical blades or spring teeth that pull thatch from the turf. However, there is a risk that you will pull out some of the turf with the thatch.

You can also decrease thatch by making the soil more hospitable to organisms that will help the thatch to decompose—such as earthworms. By adding lime to neutralize acidic soil, you encourage earthworm activity and thus aerate the soil.

Mowing

If you cut turfgrass too short, the lawn becomes sparse, inviting problems with weeds. Likewise, lawns that grow too tall between mowings can become weedy—not to mention shaggy.

Mow frequently at the heights recommended in the following table.

Recommended mowing heights

Grass species	Desired grass height (inches)*
Fine-leaf fescues	2 to 2.5
Tall fescue	2 to 3
Kentucky bluegrass	2 to 2.5
Perennial ryegrass	2 to 2.5
Zoysiagrass	1 to 1.5

*If grass is stressed by heat, drought, weeds, or other problems, aim for the higher grass height. Under normal conditions, aim for the lower height.

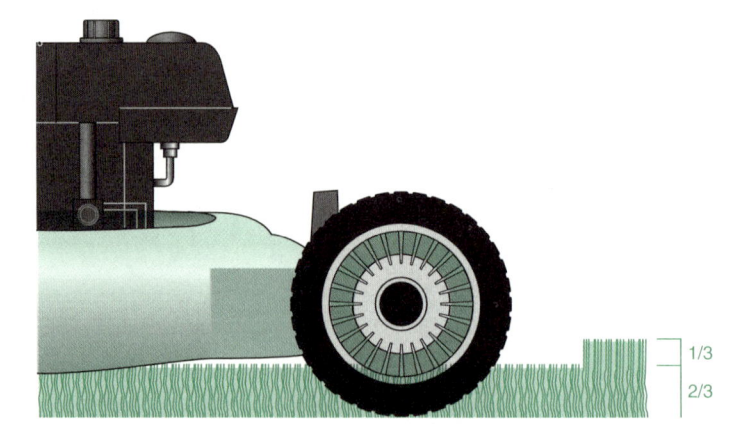

For a healthy lawn, mow at recommended heights, and never remove more than one-third of the grass blade at any one mowing.

Watering

Too much or too little watering can cause problems. An overwatered lawn can result in more mowing and clippings, as well as excessive thatch; it also encourages a shallow turf root system, which is especially vulnerable to weed and insect invasions. An underwatered lawn becomes sparse, opening the way to weed invasions.

Water lawns deeply (to the depth of the root system) and infrequently (on an "as needed" basis) at a rate no faster than the soil can absorb. To maintain green, actively growing turfgrass throughout the entire season, most lawns require 1 to 1½ inches of water per week—no matter whether the water is coming from rain or irrigation. If you water early in the day, you reduce evaporation, thereby conserving water and reducing the risk of disease problems.

If water is restricted, soil preparation and variety selection become even more critical; select a drought-tolerant grass, such as tall fescue or zoysiagrass.

Physical or mechanical removal of pests

In turf, the mechanical removal of pests simply means removing them by hand. Small populations of weeds can be controlled by pulling, digging, or hoeing. For large weed populations, physical removal is laborious, time-consuming, and often impractical.

The physical removal of insect or disease pests is generally not practiced in lawns.

Prevention and sanitation

When planting a lawn, use seed and vegetative propagules (sod, plugs, or sprigs) that are as pest-free as possible. Obviously, planting a pest-free lawn will minimize future problems. So inspect seed labels or vegetative propagules to find out if weeds are present.

If weeds are in flower, you can reduce the spread of weed seeds by collecting grass clippings when mowing established turf. Collecting clippings will also reduce the spread of some diseases during periods of infection. However, both of these strategies may have only a minimal effect on weeds and diseases.

To prevent the spread of weed seeds from areas adjacent to established turf, mow weeds in these areas before you observe flowers producing seeds.

Alternatives to pesticides

If you wind up with a pest problem, you don't have to reach automatically for a traditional, synthetic (human-made) pesticide. There are alternatives, which are generally (but not always) less toxic:

Microbial insecticides. These products combat insects with microscopic living organisms—viruses, bacteria, fungi, protozoa, and nematodes.

Botanical insecticides. These products are naturally occurring insecticides derived from plants. For instance, pyrethrins are compounds derived from the pyrethrum daisy; upon contact, they rapidly paralyze insects.

Insecticidal soaps. These soaps are formulated specifically for their ability to control insects.

CHECK IT OUT

For more information on microbials, botanicals, and insecticidal soaps: pages 56, 62.

Pesticides

Pesticides should be applied only when other corrective methods cannot produce the desired quality of turf. Here are some ways to reduce the risk from pesticides applied to lawns:

- Select pesticides that target specific pests, do not remain in the environment for a long time, and are of low toxicity.
- Make spot applications only to areas where pests are a problem rather than broadcast applications on an entire lawn.
- Read, understand, and follow label directions for the safest and most effective use.
- Check lawns for insects, and apply insecticides only when pests reach a level that can cause damage.
- Turf fungicides are expensive and require considerable experience to use properly for controlling diseases. Therefore, using fungicides on home lawns is not recommended. If you cannot control diseases with proper management, it is often cheaper to replant than to use fungicides.

Prepared with Thomas Voigt, turfgrass specialist, and Philip Nixon, entomologist, UI Cooperative Extension Service.

Additional source: Effective Lawn Care with Reduced Pesticide and Fertilizer Use, *Fact Sheet 637, Cooperative Extension Service, University of Maryland System.*

COMING TO TERMS WITH PESTS (AND IPM)

For some homeowners, the words *integrated pest management* may sound like nothing more than scientific jargon. But if you look beyond the terminology, you will discover that integrated pest management, or IPM, has a lot to offer in the home, yard, and garden.

What makes IPM unique is that it does not aim to eliminate all pests. Instead, its goals are twofold:

- To prevent or halt pest damage by keeping pest populations below damaging levels.
- To protect environmental quality, human health, and your checkbook by reducing unnecessary pesticide use.

IPM emphasizes preventive measures, which can include anything from using mulches and planting disease-resistant varieties to maintaining plants with good horticultural care.

When it is too late for prevention, IPM calls for wise pest management. Management techniques may include pruning out infested limbs or using less toxic alternatives to pesticides (or a combination of techniques) to reduce pests to nondamaging levels. Pesticides are a part of IPM when used sparingly and wisely.

Plant health care, a commercial version of IPM, is commonly used by tree-care specialists. Plant health care programs focus on keeping trees healthy as the major preventive strategy. But whatever your terminology of choice may be, the key point is that there are a variety of ways to thwart insects and diseases.

Prepared with John Lloyd, entomologist, UI Cooperative Extension Service.

2 Cut the grass without gas

Time to try something new

Whether you are in the market for a new lawn mower or are buying your first one, recent advances in lawn mower technology make this an ideal time to consider switching to an electric or an old-fashioned reel mower.

The benefits of going electric

According to the U.S. Environmental Protection Agency (EPA), electric mowers are half as loud as gas mowers, use 70 percent less energy than gas mowers, and require no maintenance, oil, gas, or tunc-ups. As an added bonus for weary backs, electric mowers start at the push of a button. You don't even have to deal with electric cords—if you prefer to avoid them—because many electric mowers come cordless.

Electric mowers use less energy and create less pollution than gas-powered mowers. Also, with today's battery-powered models, you don't have to deal with power cords.

In addition, most electric mowers cost less than 10 cents to recharge. Riding electric mowers can cut up to 1 acre on a single charge, whereas push electric mowers can cut up to ½ acre on a single charge.

Gas-powered lawn mowers, in contrast, are one of the worst unregulated sources of air pollution. Using a gas-powered lawn mower for one hour releases as many hydrocarbons into the atmosphere as driving a car 50 miles. (Hydrocarbons are air pollutants created when fossil fuels are burned.)

The U.S. EPA estimates that lawn mowers in the United States emit about 124,000 tons of hydrocarbons, 980,000 tons of carbon monoxide, and 4,200 tons of nitrogen oxide gases into the atmosphere each year. Many homeowners also dispose of oil improperly, further damaging the environment.

Your options

Electric mowers are available in either push or riding varieties, and as discharge or mulching mowers. In addition, there are three basic designs of cordless mowers: 12, 24, and 36 volt.

The 12-volt models take about a day to completely recharge. The 24-volt models have a built-in recharger and can be completely recharged in about sixteen hours. The 36-volt models, which are riding mowers, charge in about 8½ hours. Unfortunately, though, no manufacturer is currently producing new electric riding mowers. Reconditioned ones are

Because electric mowers do not contain gas, they can be folded up and turned on their side for compact storage.

available from: Norman Marcotte, P.O. Box 67, Shelburne, VT 05482, (802)985-2358; and Kansas Wind Power, 13569 214th Road, Department 57, Holton, KS 66436, (913)364-4407.

Electric lawn mowers—with or without cords, as well as discharge or mulching varieties—are available from Black and Decker, (800)762-6672, and Ryobi America, (800)345-8746. They are marketed by numerous home-and-garden and hardware stores, including Ace, Builders Square, Home Base, Home Depot, Home Quarters, Lowe's, and Sears.

What about the reel mowers?

Good old-fashioned reel mowers—push mowers without an engine—are still relatively easy to find. You can purchase them from mail-order catalogs such as Real Goods, (800)762-7325, or from many home-and-garden and hardware stores.

Prepared with Nicholas Smith-Sebasto, environmental education specialist, UI Cooperative Extension Service.

3 Prevent insect, disease, and weed problems in the garden

Your best strategies

No garden this side of Eden is going to be entirely pest-free. But you can reduce the risk of insect, disease, and weed problems with some basic management practices. A reduced risk of pest problems means a reduced use of pesticides.

Some key management recommendations follow.

Choose a sunny location

Eight to ten hours of sunlight per day are necessary for the proper growth, flowering, and fruiting of most vegetable crops. Growing vegetable plants in a sunny location speeds the evaporation of water from plant surfaces and slows the spread or buildup of disease organisms.

Select pest-resistant varieties

Some crop varieties have a built-in resistance to disease and certain insect pests. For information on pest-resistant varieties, see Chapter 4.

Start with disease-free materials

To plant disease-free seeds, gardeners are generally encouraged *not* to save their own seed but to purchase seed from reputable seed dealers.

Hot-water seed treatments help to prevent disease and are especially important with cabbage, broccoli, brussels sprouts, and cauliflower seeds. Although it is possible to do the treatment yourself, most gardeners buy seed that already has been treated by the seed producers. Many vegetable seeds also are coated with a fungicide to prevent disease—you can tell by the colored coating on the seed.

Information about the kind of treatment that the seed has received will appear on the seed package. Some companies offer untreated seed for those who wish to avoid fungicides, but you may need to specify "untreated seed" on your order.

All planting material should be healthy and free of yellowing and brown or black spots, and should not be stunted or show poor development. It never pays to buy and plant diseased transplants, no matter what the price.

Rotate crops

Certain diseases survive the winter on crop debris and may build up over time if you plant your garden in the same area each year. To prevent this buildup, do not plant any crop from the same group of vegetables in the same portion of your garden more than once every three or four years. For example, if you plant tomatoes one year, you should not plant any other crop from the tomato family in that spot of the garden for three or four years.

Rotate crops among these major vegetable groups:

- Cabbage family (Cruciferae): broccoli, brussels sprouts, cabbage, cauliflower, Chinese cabbage, mustard greens, kohlrabi, radish, rutabaga, and turnip.
- Cucumber family (Cucurbitaceae): cucumber, gourds, muskmelon, pumpkin, all types of squash, and watermelon.
- Tomato family (Solanaceae): eggplant, husk tomato (ground cherry), potato, pepper, and tomato.
- Goosefoot family (Chenopodiaceae): red beet and spinach.
- Onion family (Amaryllidaceae): chives, leeks, onions, shallots, and garlic.
- Pea family (Leguminosae): peas and all types of beans.
- Carrot family (Umbelliferae): carrot, parsnip, and celery.

Provide an adequately fertile soil

Properly fertilized plants are generally less susceptible to disease than those that are underfertilized or overfertilized.

To determine the amounts and availability of nutrients in your garden, have the soil tested. Gather small amounts of soil from seven or eight well-scattered locations in your garden, mix the soil together, dry at room temperature, and seal in a sturdy $1/2$-pint container. Write "for vegetable garden" on the container, along with your name and address, and send it to the nearest soil-testing laboratory. In a few weeks, you will receive the results of the test as well as fertilizer and lime recommendations for your garden.

> **✓ CHECK IT OUT**
> For information on alternatives to traditional, synthetic fertilizers: page 81.

Maintain the proper soil pH

A soil that is slightly acidic to neutral is best for growing most vegetables. This means the soil pH should be anywhere from 6.0 to 6.9. (A pH of 7.0 is considered neutral; anything below 7.0 is acidic, and anything above is alkaline.)

If your soil test indicates that the soil is more acidic than it should be (a pH below 6.0), apply the recommended amount of limestone. Lime raises the pH, but be sure to avoid overliming. If your soil is too alkaline (pH above 7.5), add sulfur.

Work the lime or sulfur into the soil at the same time that you apply fertilizer. A single application of lime or sulfur is usually adequate for four to five years. After that, have the soil retested before making additional applications.

Provide good air circulation

Planting vegetables at recommended spacings allows the air and sun to dry off the lower foliage and therefore reduces the relative humidity around the plants. Staking, caging, trellising, and pruning plants also provide more air circulation.

Plant in well-drained soil

In heavy or poorly drained soils, try planting on ridges, hills, or in raised beds to prevent seedling blights, root rots, or foliage diseases in plants that contact damp soil. Adding organic matter to the soil will help to loosen poorly drained clay soils.

Observe recommended planting dates

Plant seeds in soil that is warm enough for rapid germination. If you wait to sow seeds until the soil has warmed, young plants grow faster and often escape infection. For the optimum planting dates, check seed packets, gardening references, and Cooperative Extension Service guidelines.

CHECK IT OUT

For information on using beneficial insects to control pests: page 67. For information on using barriers or "solarizing" to reduce pest problems: pages 77, 79.

Water properly

If plants suffer from too much or too little water, they will be less vigorous and more susceptible to problems. Water the soil thoroughly by adding an inch of water per week when there is not sufficient rainfall.

Trickle irrigation and watering with soaker hoses are good alternatives to overhead irrigation with a sprinkler. With trickle irrigation and soaker hoses, water trickles out along the hose, usually at low pressure. These watering techniques do not wet the foliage, thereby reducing the risk of diseases that attack foliage.

If you use an overhead sprinkler, water early in the day so the foliage can dry before nightfall.

Destroy crop residues after harvest

If diseases have been a problem, it is best to remove the plants after harvesting each crop; then compost or burn them. This will prevent the disease from overwintering on the debris and being carried over to the next season.

Prepared with Darin Eastburn, plant pathologist, and James Schmidt, horticulturist, UI Cooperative Extension Service.

4 Identify fruits and vegetables that need fewer pesticides

Selecting the most resistant crops

Although some crops suffer less insect and disease damage than others, virtually all fruits and vegetables are attacked by insects and pathogens (disease-causing organisms). Fortunately, high yields and perfect quality are not essential in home gardens. As a result, you can grow several crops that require minimal effort in controlling insects and diseases. The accompanying table lists examples of such crops.

Crops that require minimal pest control

Asparagus	Peas
Beets	Peppers
Blackberries	Radishes
Blueberries	Raspberries
Currants	Strawberries
Herbs, many varieties	Sweet corn, early season
Leafy greens (leaf lettuce, spinach, and others)	Sweet potatoes
	Tomatoes (VF varieties)
Onions	

If the crops listed in the table above are started from disease-free seeds or transplants, and if they are grown on well-drained soils, midwestern gardeners can produce acceptable crops without repeated pesticide applications. Some losses are likely, but so are successful harvests. In the warmer climate of southern states, losses are often greater.

Other crops often require more protection with pesticides and are frequently difficult to grow in the Midwest.

Crops that require more pest control

Crop	Possible pests
Apples	Several insects, diseases
Cucumbers, muskmelons	Cucumber beetles, bacterial wilt
Squash	Squash bug, squash vine borer
Sweet corn, late season	Corn earworm, European corn borer, fall armyworm

Losses will differ somewhat each season, and so will the amount of pest control needed. One way to reduce exposure to pesticides is to spray only the crops in your garden that have pest problems rather than spraying the entire garden. Insect and disease problems generally are limited to one type of plant; they don't spread to different types of plants in the garden.

Selecting resistant varieties

Specific varieties of some crops often differ significantly in their ability to withstand attack by insect pests and especially by plant diseases. Apple varieties, for example, vary considerably in their susceptibility to apple scab, a common and sometimes devastating fungal disease. Very susceptible varieties include Red Delicious, Golden Delicious, McIntosh, Winesap, and Jonathan. Those that are highly resistant to the scab fungus include recently developed varieties such as Prima, Priscilla, Jonafree, Freedom, Liberty, Dayton, Williams Pride, Goldrush, and Enterprise.

Although the availability of *disease-resistant* fruits and vegetables is widely known, *insect-resistant* plant varieties have been more difficult to develop. Therefore, examples of insect resistance in vegetables are few. Butternut squash is more resistant to the squash vine borer than are either acorn squash or Blue Hubbard squash. And among summer squash, zucchini is the most susceptible to damage by squash vine borer.

Hybrid varieties usually offer the best tolerance to diseases and insects. A hybrid, by definition, results from crossing (breeding) two parental lines that differ in one or more important characteristics. Commercial hybrids combine such desirable characteristics as uniformity of plant and fruit type, uniform maturity, disease tolerance, improved quality, and increased overall vigor. Hybrid plants are usually stronger, healthier, and more productive than open-pollinated varieties. They may tolerate air pollution and weather conditions better and often bloom earlier.

For more information on disease-resistant vegetable and fruit varieties in the Midwest, send a written request for a copy of *Report on Plant Diseases #900* and $1 to the Department of Crop Sciences, University of Illinois, N-533 Turner Hall, 1102 S. Goodwin Ave., Urbana, IL 61801. Make checks payable to the University of Illinois.

Prepared with Stephen Ries, plant pathologist, and Rick Weinzierl, entomologist, UI Cooperative Extension Service.

5. Select native trees and shrubs that will thrive in your yard

Go native

When selecting trees and shrubs for your yard, you will want species that are attractive, serve a special function (provide shade, for instance), or both. But before you can even consider such qualities, your first task is to narrow down the list to trees and shrubs that are well adapted to your environment. After all, an attractive tree serves little purpose if it will not survive.

The key to finding woody plants that will thrive in your environment is to focus on *native* trees and shrubs. Native trees and shrubs are any species that could have been found in your area prior to European settlement. They are well adapted to your region's climate. Realize, however, that even native plants will undergo more stress in a residential landscape than in their natural habitat, such as a forest.

> ✓ **CHECK IT OUT**
> For more information on native plants: page 33.

As you consider native trees and shrubs that will thrive in your area, the major considerations are:

Drought tolerance. Because the urban environment is warmer and drier than the open countryside, drought tolerance is especially important. Trees with leathery leaves and vigorous roots are often drought-tolerant.

Soil drainage and alkalinity. Some trees and shrubs do well in poorly drained soils, whereas others require well-drained soil. Also, some trees and shrubs are sensitive to alkaline (high pH) soils. Therefore, consult your Natural Resources Conservation Service (formerly the USDA Soil Conservation Service) for soil information in your area. Then select trees and shrubs accordingly.

Pest susceptibility. No tree or shrub species is entirely free of insect and disease problems, but some species are less susceptible than others. Stressful conditions can predispose a plant to pest infestation; therefore, the tolerance of a species to drought, poor drainage, and alkaline soil can affect its susceptibility to pests.

The following chart rates some common trees and shrubs on these key qualities. The species listed are native to many areas throughout the Midwest and Northeast—which explains why the evergreen list is so short. Most evergreens grown in the United States are native to the West.

Native trees and shrubs in the Midwest and Northeast

Deciduous trees

	Tolerance to drought	Tolerance to poor drainage	Tolerance to alkaline soils	Ease of transplanting	Pest resistance*
Ash, white, *Fraxinus americana*	0	0	+	0	
Bald cypress, *Taxodium distichum*	-	+	0	0	i,d
Beech, American, *Fagus grandifolia*	-	-	-	-	i,d
Birch, river or red, *Betula nigra*	-	+	-	0	i,d
Buckeye, Ohio or fetid, *Aesculus glabra*	0	0	0	0	i
Coffeetree, Kentucky, *Gymnocladus dioica*	+	0	+	0	i,d
Dogwood, flowering, *Cornus florida*	-	-	0	-	
Hackberry, sugar, *Celtis laevigata*	+	+	+	+	i,d
Hawthorn, cockspur, *Crataegus crus-galli*	+	0	+	0	
Holly, American, *Ilex opaca*	0	0	0	-	
Honeylocust, thornless, *Gleditsia triacanthos* var. *inermis*	+	+	+	+	
Hornbeam, American (Blue Beech), *Carpinus caroliniana*	-	-	0	-	i
Ironwood (hophornbeam), *Ostrya virginiana*	0	0	0	-	i,d
Magnolia, cucumbertree, *Magnolia acuminata*	-	-	0	-	i,d
Maple, black, *Acer nigrum*	+	-	+	0	i

(See table key on page 26.)

	Tolerance to drought	Tolerance to poor drainage	Tolerance to alkaline soils	Ease of transplanting	Pest resistance*
Maple, red or swamp, *Acer rubrum*	-	+	-	0	
Maple, sugar or rock, *Acer saccharum*	0	-	+	0	
Oak, bur, *Quercus macrocarpa*	+	0	+	0	i
Oak, pin or swamp, *Quercus palustris*	0	+	-	+	i
Oak, red, *Quercus rubra*	0	-	-	-	i
Oak, Schumard, *Quercus shumardii*	+	0	+	0	i,d
Oak, swamp white, *Quercus bicolor*	0	+	0	0	i,d
Oak, white, *Quercus alba*	0	-	-	0	i
Redbud, *Cercis canadensis*	+	-	+	0	i
Viburnum, blackhaw, *Viburnum prunifolium*	0	0	0	0	
Yellowwood, *Cladrastis lutea*	0	0	+	0	i
Evergreen trees					
Arborvitae, white cedar, *Thuja occidentalis*	-	+	+	0	
Pine, eastern white, *Pinus strobus*	0	-	0	0	
Deciduous shrubs					
Allegheny serviceberry, *Amelanchier laevis*	+	0	0	+	
American cranberrybush viburnum, *Viburnum trilobum*	0	0	0	+	i,d

Native trees and shrubs in the Midwest and Northeast, cont.
Deciduous shrubs, cont.

	Tolerance to drought	Tolerance to poor drainage	Tolerance to alkaline soils	Ease of transplanting	Pest resistance*
American elder, *Sambucus canadensis*	+	0	0	+	i
Arrowwood viburnum, *Viburnum dentatum*	-	+	0	+	i,d
Beach plum, *Prunus maritima*	+	-	0	+	
Black chokeberry, *Aronia melanocarpa*	+	+	0	+	i,d
Blackhaw viburnum, *Viburnum prunifolium*	+	+	0	+	i,d
Bottlebrush buckeye, *Aesclus parviflora*	0	0	0	0	i,d
Brook euonymus, *Euonymus americanus*	0	0	0	+	
Checkerberry, *Gaultheria procumbens*	0	0	-	+	i,d
Common ninebark, *Physocarpus intermedius opulifolius*	+	+	+	+	i,d
Common sweetshrub, *Calycanthus floridus*	0	0	+	+	i
Common winterberry, *Ilex verticillata*	+	0	-	+	i
Common witchhazel, *Hamamelis virginiana*	-	+	0	0	i,d
Dwarf fothergilla, *Fothergilla gardenii*	0	-	-	0	i,d
Flame azalea, *Rhododendron calendulaceum*	+	0	-	-	
Flame leaf (shining) sumac, *Rhus copallina*	+	0	+	+	i,d

(See table key on page 26.)

	Tolerance to drought	Tolerance to poor drainage	Tolerance to alkaline soils	Ease of transplanting	Pest resistance*
Golden weeping willow, *Salix tristis*	-	+	-	+	
Gray dogwood, *Cornus racemosa*	+	+	0	+	i,d
Highbush blueberry, *Vaccinium corymbosum*	-	+	-	0	i
Inkberry, *Ilex glabra*	-	+	-	+	i,d
Large fothergilla, *Fothergilla major*	0	-	-	0	i,d
Mapleleaf viburnum, *Viburnum acerifolium*	-	+	-	0	i,d
New Jersey tea, *Ceanothus americanus*	+	-	0	-	i
Northern bayberry, *Myrica pensylvanica*	+	+	-	+	i,d
Oakleaf hydrangea, *Hydrangea quercifolia*	-	0	0	+	i
Red chokeberry, *Aronia arbutifolia*	+	+	0	+	i,d
Redosier dogwood, *Cornus stolonifera (sericea)*	0	+	0	+	
Shadblow serviceberry, *Amelanchier canadensis*	+	0	0	+	d
Smooth sumac, *Rhus glabra*	0	-	0	+	i,d
Staghorn sumac, *Rhus typhina*	+	-	0	+	i
Summersweet clethra, *Clethra alnifolia*	0	+	-	+	i,d
Virginia rose, *Rosa virginiana*	+	0	0	+	
Western sandcherry, *Prunus besseyi*	+	0	0	+	

Native trees and shrubs in the Midwest and Northeast, cont.

Deciduous shrubs, cont.

	Tolerance to drought	Tolerance to poor drainage	Tolerance to alkaline soils	Ease of transplanting	Pest resistance*
White fringetree, *Chionanthus virginicus*	0	-	-	-	i,d
Witherod viburnum, *Viburnum cassinoides*	0	0	0	+	i,d

Evergreen shrubs

Canby paxistima, *Paxistima canbyi*	0	-	+	+	i
Drooping leucothoe, *Leucothoe fontanesiana (catesbaei)*	-	+	-	+	i
Eastern red cedar cultivars, *Juniperus virginiana* cultivars	+	-	+	+	
Mountain-laurel kalmia, *Kalmia latifolia*	-	-	-	+	i

SOURCE: Information on tree varieties is based in part on *Selecting and Planting Trees*, The Morton Aboretum, Lisle, Illinois, 1990. Information on shrub varieties is based in part on *Native Trees, Shrubs, and Vines for Urban and Rural America: A Planting Design Manual for Environmental Designers*, by Gary L. Hightshoe, Iowa State University, 1988.

Table key

Tolerance to...
+ = tolerant
0 = intermediate or uncertain
- = intolerant

Ease of transplanting
+ = easy to transplant
0 = intermediate
- = difficult to transplant

Pest resistance
d = relatively disease-free
i = relatively insect-free
Trees without the "d" do not have any special resistance to disease.
Trees without the "i" have no special resistance to insects.

*No plant is totally resistant to pest problems. The plants listed with a "d" or "i" are less likely to have pest problems.

Prepared with David Williams, woody ornamentals specialist, John Lloyd, entomologist, Nancy Pataky, plant pathologist, and Philip Nixon, entomologist, UI Cooperative Extension Service.

6 Keep your trees healthy and safe

Live long and prosper

Many trees have life spans that any human would envy. But in residential areas, these impressive life spans are often cut short by the rigors of the urban environment.

With attention to a few key management practices, this doesn't have to be so. Pay particular attention to watering, fertilization, and pruning.

Watering

Proper watering is critical when transplanting a new tree. Water regularly, but be careful not to overwater. The best way to know if the roots need water is to check the soil 2 inches into the ground near the root ball. If the soil forms a ball when you squeeze it, moisture is adequate; if the soil crumbles, it is too dry.

After a tree is established, normal rainfall usually will provide an adequate amount of water. But you still need to water during droughts or extended dry periods during the summer—when less than an inch of rain has fallen in a two-week period and temperatures have passed 85°. During these periods, water generously as needed.

Fertilization

Once a tree is established, apply up to 6 pounds of nitrogen per 1,000 square feet of soil surface each year to maximize tree growth. Every 3 years, apply 3.6 pounds of phosphorus and 6 pounds of potash per 1,000 square feet.

Apply fertilizer to an area that extends several feet beyond the canopy drip line (see illustration, page 28). Research has shown that a tree's roots may extend as far out as two to three times the height of the tree. The following are two common application strategies:

Surface broadcasting. You can broadcast dry fertilizers on the soil surface around the tree by hand or with an ordinary lawn fertilizer spreader.

Beware, however, that broadcasting the entire amount of recommended nitrogen at one time will burn the lawn. Therefore, do not apply more than 3 pounds of actual nitrogen per 1,000 square feet at any one time. Also, be sure to soak the fertilizer into the soil to reduce the risk of injuring the lawn.

Placing fertilizer in holes. You can apply the full rate of nitrogen (6 pounds per 1,000 square feet) at one time if you place the dry fertilizer in holes. Using a punch bar or a drill with a 2-inch auger, create holes beginning about 2 feet from the tree trunk. Place the holes in a grid pattern 2 to 3 feet apart (see illustration). The holes should be 12 to 18 inches deep and at a slight angle. Pour fertilizer directly into these holes. The amount you should place in each hole will vary according to the fertilizer formulation used, so consult your local Cooperative Extension Service or garden center professional for the proper application rates.

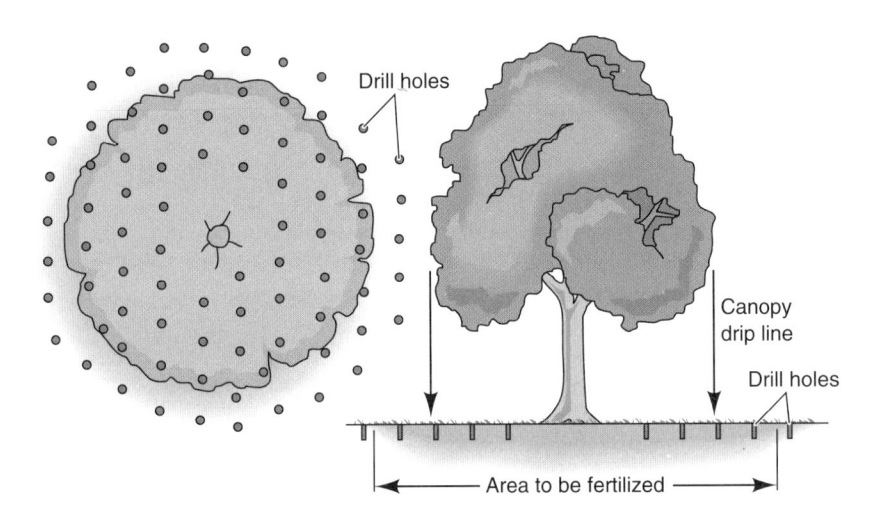

Fertilize to an area that extends several feet beyond the canopy drip line. By placing fertilizer in holes, you can apply the full rate of nitrogen at one time— 6 pounds per 1,000 square feet.

Pruning

If you select the right tree and plant it in an area where it will be clear of buildings and walkways, the need for pruning will be minimal. Once you achieve good branch structure, you will only need to remove dead wood periodically.

Warning signs

Be on the lookout for these warning signs, which may indicate that your tree is under stress:

Leaves. Are they off-color, undersized, withering, or showing dark blotches? Are many of them missing?

Branches. Are the branches dying at the ends? Also, do the branches interfere with wires, gutters, chimneys, or windows? How much have the tips of the branches grown in the past year? You can tell by looking at the "bud scale scars" (see illustration). In general, twig growth on most

- Terminal bud
- Lateral buds
- Leaf scars
- Terminal bud scar

Latest year's growth

Previous year's growth

- Terminal bud scar

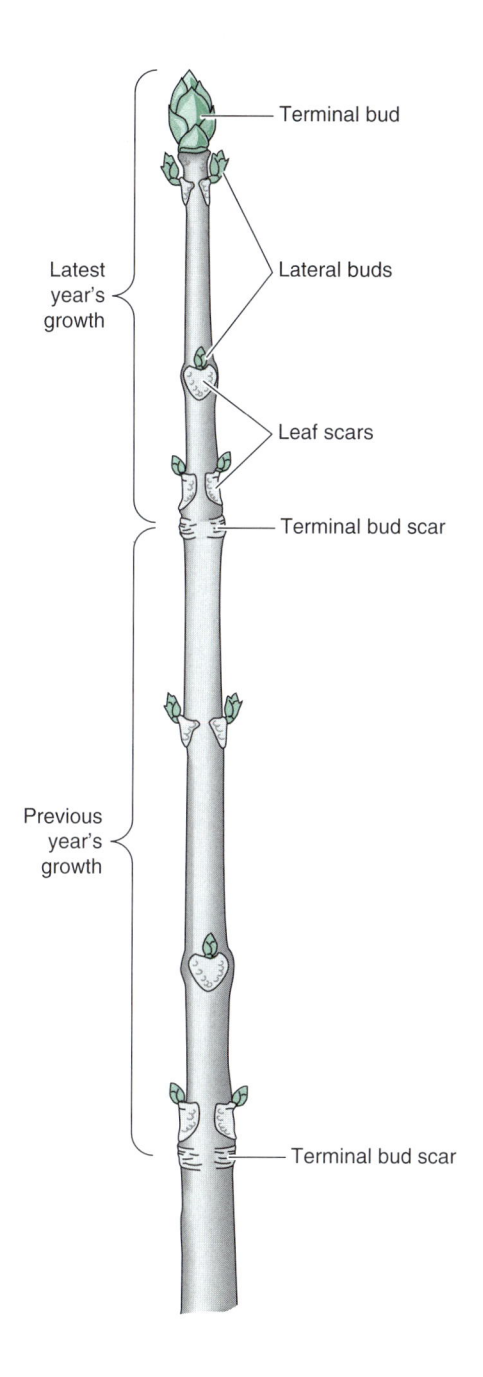

Look at the "terminal bud scars" for clues to the health of your tree. The distance between one terminal bud scar and the next shows how much a tree branch has grown during one year. Little growth in the branches may be a sign that the tree is under stress.

trees should be at least 9 inches per year. Trees approaching mature size may grow only 6 to 9 inches per year.

Trunk. Are there cracks or holes in the tree trunk or cankers (localized dead areas) present on the bark?

Insects. Do you see insect activity or signs of insect presence: curled leaves, chewed leaves, missing leaves, holes in branches, buds that don't sprout, webs, sawdust, holes in the trunks, or galls (tumors that form on plant tissue)?

Storms. Has a storm broken limbs or done any other damage?

Soil. Is the soil compacted, or packed down? Compaction, one of the major killers of trees in urban areas, reduces soil aeration—the availability of air in the soil. This can suffocate the tree because roots need air to carry out respiration and take in nutrients.

If you suspect a compaction problem, aerate the soil. To aerate, drill 2-inch holes at 2-feet intervals in concentric circles around the base of the tree, starting 3 feet from the trunk and extending just beyond the drip line. Fill the holes with fine gravel or coarse sand to permit continued aeration of the root system.

Some of these warning signs can have a number of causes, so you may need to call in a professional arborist to determine the problem. For instance, yellowed, undersized, or withered leaves can be caused by diseases, insects, inadequate moisture, or lack of nutrients.

Construction damage

Some types of construction damage to tree limbs and trunks are obvious. But some construction-related damage is not evident until later, when foliage may wilt or become off-color, bark becomes loosened, suckers appear at the base or on the trunk of the tree, or branches start to die back.

The amount of damage, the species of tree involved, and the soil type all determine how long it will take symptoms to appear.

Controlling construction damage

The following are ways to limit damage caused by construction:

- Reduce traffic as much as possible around the construction site. Work with your contractor to establish definite traffic lanes and fence them off if necessary.
- If construction traffic must pass close to trees, use "bridging" to protect roots. To do this, elevate a metal or heavy wood walkway on railroad ties.
- Locate stockpile areas for soil and building materials well away from the drip line of trees you want to save.

- When you install temporary or permanent driveways or traffic lanes, cut nearby tree roots cleanly. Cleanly cut roots will heal well, and new roots will develop.
- If trees have lost some roots or are in compacted soil due to construction, they usually need water on a regular basis.
- If you must prune a tree's roots, also remove a portion of the top of the tree to maintain the same ratio between roots and shoots.
- If not done properly, changing the grade of the soil over tree roots can damage the tree. Raising the soil level over the roots can cut off the tree's air supply, smothering it. Lowering the soil level can cause loss of roots. For details on how to prevent this and other construction damage, obtain a copy of *Pruning and Care of Trees and Shrubs (U5040)*. This 60-page guide is available by writing to: Vocational Agriculture Service, University of Illinois, College of Agricultural, Consumer and Environmental Sciences, 1401 S. Maryland Dr., Urbana, IL 61801. Telephone: (217)333-3871.
- When adding soil, be sure the fill soil is as porous as possible. Sandy soil is preferred because it permits natural drainage of water and does not easily compact. Two or 3 inches of sandy soil can be filled over a root system without harming the tree, but 2 to 3 inches of clay soil over the roots will kill some tree species.

Prepared with David Williams, woody ornamentals specialist, Floyd Giles, horticulturist, and Patrick Weicherding, urban forestry specialist, UI Cooperative Extension Service.

Additional source: How to Hire an Arborist, Tree City USA Bulletin No. 6, National Arbor Day Foundation, 100 Arbor Ave., Nebraska City, NE 68410.

IS TREATED WOOD SAFE?

Lumber is often treated to protect it from decay by fungi and attack by insects. But some consumer advocates are concerned that treated wood poses health and environmental hazards, especially when used for playground equipment, picnic tables, and decks. The potential risk depends on what kind of wood preservative you're talking about.

- **CCA:** Most of the treated wood you find around the yard and in the home is treated with CCA—a "waterborne arsenical preservative." But there is no evidence that it poses a serious environmental or health risk. CCA-treated wood can be used for decks, play structures, and picnic tables.

- **Creosote:** The main place where you might find creosote-treated wood around the home is on railroad ties, used to create garden beds. Older railroad ties usually pose little problem. But if you use a newer one with fresh creosote on it, the creosote will burn any vegetation that it contacts. If fresh creosote is washed from the railroad tie, it can also burn plants up to about a foot away.

- **Pentachlorophenol:** Penta is probably the most effective wood preservative, especially against insects. But because of its potency and volatility, it is usually not used around the yard; any plant tissue that the fumes contact will be killed. Penta should never be used inside the house where plants are to be grown.

These three types of preservatives are considered "restricted-use" products because of the hazards they pose to the people applying them. Restricted-use status means they can only be sold to and used by certified pesticide applicators or those working under their supervision. It means the wood preservatives can be hazardous to people using them—not necessarily that the treated wood itself poses great risk.

Although CCA-treated wood is fine for picnic tables, do not use it as a countertop or chopping board. Why the difference? Picnic tables are mainly used for serving prepared foods, whereas a countertop and chopping board are used for preparing raw food.

If you want to take extra precautions, use a clear varnish to cover all treated wood with which you may come in contact. Not only will the varnish serve as a barrier between you and the treated wood, but it also will extend the wood's life.

Also, never burn treated wood in a fireplace or wood-burning stove.

Prepared with Poo Chow, professor of wood science, University of Illinois, and Floyd Giles, horticulturist, UI Cooperative Extension Service.

7 Plant 'wild' in your backyard

A private wilderness

They have been called "natural gardens"—patches of prairie scattered along railroad tracks or in old cemeteries, echoes of pioneer days when an inland sea of waving grass and colorful wildflowers covered most of the central United States. If you would like to add a similar touch of the wild to your own backyard, try creating your own private wilderness.

The key is using plants that are native to your area. The phrase *native plants* is used in different ways, but for the purposes of this publication, a native plant is any species that could have been found in your area prior to European settlement. It is a broad term that includes native grasses, wildflowers, trees, and shrubs.

Native plants offer these advantages:
- They are well adapted to your region's climate.
- They require less maintenance after they are established.
- They provide good cover, nesting sites, and food for certain desirable wildlife.
- They are attractive, unique alternatives to traditional landscape designs.

Before you create a backyard wilderness, however, check with your municipal officials to find out if any local ordinances would prohibit your plans; some communities put restrictions on how high certain types of plants may be allowed to grow.

Plant selection

As you select native plants for your backyard wilderness, take into consideration several factors—the climate, the soil and topography, the amount of shade or sun, and the importance of plant diversity.

Climate. The U.S. Department of Agriculture has divided North America into eleven zones of plant hardiness. Horticulturists rate trees, shrubs, and perennial flowering plants according to the northernmost plant hardiness zone in which they will normally survive the winter (see map). Make sure you select plants that will grow in your zone.

Soil and topography. Different plants also grow best in certain soils. To learn about your soil, contact your nearest Cooperative Extension Service or Natural Resources Conservation Service (NRCS) office. Specialists can help you determine whether your soil is acidic or alkaline; clay, loam, silt, or sand; deficient in nutrients; well drained or poorly drained.

Range of average annual minimum temperatures

Zone 1 (below −50°F)
Zone 2 (−50° to −40°F)
Zone 3 (−40° to −30°F)
Zone 4 (−30° to −20°F)
Zone 5 (−20° to −10°F)
Zone 6 (−10° to 0°F)
Zone 7 (0° to +10°F)
Zone 8 (+10° to +20°F)
Zone 9 (+20° to +30°F)
Zone 10 (+30° to +40°F)
Zone 11 (above +40°F)

Plant hardiness zones in the continental U.S.

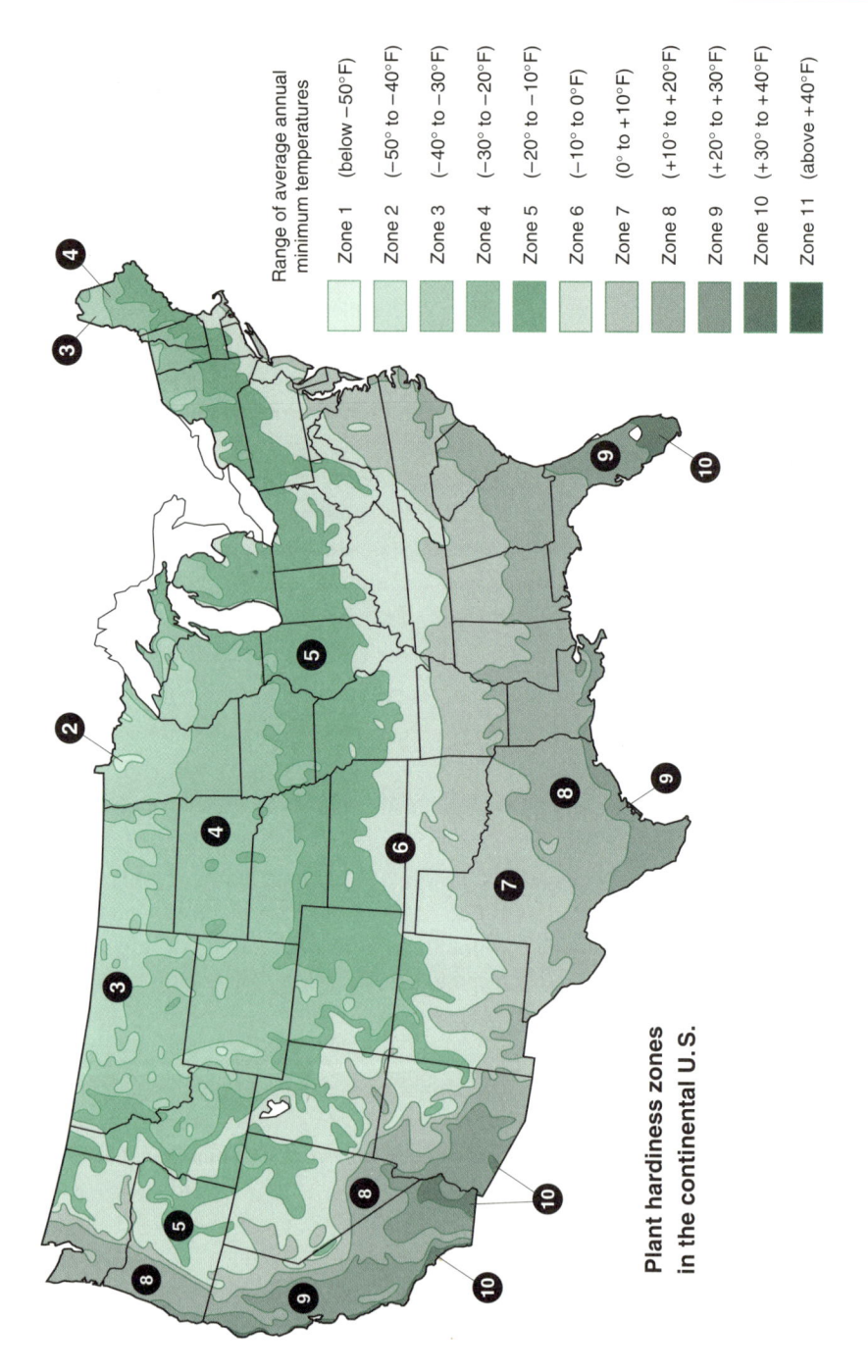

This map is not large enough to show the short, thin strips of zone 11 that can be found in a few spots along the southwestern edge of California. Most of the Hawaiian islands, not pictured here, fall in zone 11, although the central portion of Hawaii and the north-central part of Kaui are in zone 10. Zone 1 also does not show up on the map above, but it can be found in portions of Alaska. Alaska ranges from zone 1 to zone 8.

The amount of shade or sun. In most cases, the shade or lack of it in your backyard will limit your selection of native plants. If your yard has plenty of both, consider a shaded woodland garden *and* a sunny prairie garden. By doing so, you are more likely to have plants in bloom throughout the growing season.

Plant diversity. When a piece of land is left undisturbed, it will naturally move toward a more diverse, stable community—a process known as "ecological succession." By going with a diversity of plants in your backyard wilderness, you help to provide this same stability, and you are rewarded with a more interesting environment. Look for a healthy mix of dominant trees, small flowering trees, shrubs, wildflowers, and grasses.

Where to find native plants

To find out what plants are native to your area, contact a nearby arboretum, university or college (botany or horticulture department), your state natural resources department, or the nearest Cooperative Extension Service or NRCS office. Many garden supply catalogs and gardening books also include information on native plants.

When you are ready to collect native plants, you have several options:

Buy nursery-grown plants. This is the easiest strategy, although many nurseries offer only a limited selection—the most popular native plants. Plant in the fall or early spring so that plants are well established before the hot, dry weather of summer.

Transplant from land to be bulldozed. Do not pick or dig up native plants from the wild; the remaining prairie and woodland remnants are too precious to disturb. However, if a piece of land is scheduled to be bulldozed or plowed, you may be able to get permission to transplant native plants that would be destroyed.

Raise your own plants from seed. Do not purchase seeds from far away because they may not be well adapted to your climate. The closer the source of seeds, the better. For instance, the NRCS does not recommend buying seeds that were produced more than 200 miles north, 100 miles south, or 250 miles east or west.

Start the seeds in the fall, planting them in flats or peat pots. Use flats filled with a 50/50 mixture of sand and commercial potting mixture.

By next spring, some species will sprout, but some can take almost a year to appear. Seeds with "embryo dormancy" germinate much faster than those with "seedcoat dormancy." Seedcoat dormancy means the outer coating of the seeds is much harder and takes longer to break down.

To speed the germination process, dampen the seeds slightly, place them in plastic bags, and put them in cold storage (34°F is ideal) for about two months. Then plant them in flats or pots. The idea behind this process, known as "stratification," is for seeds to respond as if they have gone through an entire winter—germinating as if it were spring.

By the following midspring, some plants will be ready for transplanting into your backyard wilderness. Before transplanting, the roots should be well developed, and the plant should display about four or five leaves.

Ways to use native wildflowers and grasses

You can use native wildflowers to complement your ornamental plants or rock garden, attract butterflies (see Chapter 8), or create a shady woodland garden. You can grow them in containers (if space is limited), or you can join the increasing number of people in the Midwest who are replacing their traditional lawns with a backyard prairie—a mixture of native grasses and wildflowers. A backyard prairie requires less maintenance, water, fertilizer, and time than a typical lawn.

Whatever your goal, you don't have to start big but may want to slowly convert your lawn to prairie in stages. The following are some key ideas on developing a backyard prairie.

Planning your backyard prairie

To plan your own personal prairie, first be sure to find out whether local ordinances forbid such "wild" growth. Then follow these tips:

- Select a sunny site. Prairie plants require considerable sunlight.
- Avoid rows, square plantings, or pointed corners. Instead, use curves, gentle turns, and irregular plantings—similar to those found in nature.
- Place tall plants toward the back, where they will not droop over smaller plants and obscure their flowers. Tall plants can be out of place in prairie gardens of 100 square feet or less.
- Keep in mind that you are planning for three prairies—spring, summer, and fall. Make sure that spring-, summer-, and fall-blooming plants are well represented throughout your prairie garden so there is continuous color throughout the year. Spring prairie plants tend to be short (less than 2 feet high), autumn prairie plants are usually tall (6 to 8 feet high), and summer prairie plants fall in between.
- Include enough of each species so they are noticeable when in flower. Fifteen plants in an area 20 feet wide and 100 feet long will probably not provide the desired effect. For best results, space plants about 12 inches apart.
- Select a mixture of wildflowers and grasses that are well adapted to your region. (See the accompanying chart for suggested plants in the Midwest.) Grasses should make up about 60 to 90 percent of the mix, and wildflowers should make up about 10 to 40 percent of the mix.

- If you hope to attract wildlife, select a diversity of plants.
- Select plants with colors that provide contrast.
- If your site is large, consider putting paths through the prairie garden.

Establishing a backyard prairie

1. Remove grass sod before beginning cultivation. If you cannot physically remove the sod, a 1 percent solution of Roundup will kill the grass. (Use caution and follow label directions.)
2. Cultivate with a garden tiller in early spring. Tilling will bring weed seeds to the surface. Let these weed seeds germinate in the spring, then pull the weeds or kill them with herbicide. Competition from weeds can reduce the vigor of or smother your prairie plants.
3. Plant with seeds or transplants. Plants are much easier to work with than seeds, and it takes longer to establish a prairie garden using seeds—typically about three years. However, using seeds usually provides better ground cover, which means fewer open areas of weed-prone bare soil.

Maintaining a backyard prairie

A backyard prairie is relatively maintenance-free, especially after several years. But you need to monitor a few potential problems.

Weeds. Hand weeding, although slow and labor-intensive, is the most practical strategy. You may have to repeat hand weeding several times during the first year or two. Do not use any herbicides after planting or transplanting because they can kill desirable plants.

Staking. In the absence of competition, some prairie plants can grow tall and lanky. These include prairie dock, compass plant, asters, goldenrods, rattlesnake master, big bluestem grass, and Indiangrass. You will probably need to support these plants with stakes. Asters, such as New England aster, can be clipped back in July to keep them short and compact when they bloom (if desired). Also, place big grasses toward the interior of the planting, where they will be supported by other plants.

Thatch removal. On larger prairie gardens, experts advise burning grasses in March to remove dead vegetation (thatch) and to encourage new growth; the idea is to simulate the natural prairie fires of yesteryear. However, such burning is prohibited in many urban areas. Instead, you'll want to mow or use a weed whip to cut the prairie grasses quite close to the ground in early spring before plants begin to grow. Remove the thatch after you are finished.

Pests and disease. Prairie plants are remarkably free of many common insect and disease problems, and they tolerate their specific pests fairly well. It is generally best if you do not spray for undesirable insects or fungi because pesticides may destroy beneficial insects.

Mulching. When planting in the fall, use a straw, peat, or other suitable mulch. A mulch reduces "frost heaving," in which frost pushes plants out of the ground. Frost heaving can expose plants, buds, or root crowns to winter weather, killing or injuring the plant.

A prairie for all seasons
Native prairie plants in the Midwest

Late spring to early summer blooming plants

Common name	Species	Flower color	Special features	Height
Cream wild indigo (W)	*Baptisia leucophaea*	Cream	Graceful drooping appearance. Forms small mounds.	1' to 2½'
Downy phlox (W)	*Phlox pilosa*	Rose	Abundant, colorful flowers.	1' to 2'
Golden Alexanders (W)	*Zizia aurea*	Yellow	Flat clusters of tiny flowers.	1' to 3'
Hoary puccoon (W)	*Lithospermum canescens*	Golden yellow	Plant covered with stiff, whitish hairs.	8" to 1'
Prairie coreopsis (W)	*Coreopsis palmata*	Yellow	Large flowers. Shining leaves.	1½' to 3'
Shooting star (W)	*Dodecatheon meadia*	Pink	Leaves in cluster at base of plant. Flowers hang downward.	1'
Spiderwort (W)	*Tradescantia ohiensis*	Violet-blue	Smooth, pale, long, thin leaves.	2' to 3'
Wild strawberry (W)	*Fragaria virginiana*	White	Tiny, edible berries.	5" to 8"

Summer-blooming plants

Common name	Species	Flower color	Special features	Height
Big bluestem (G)	*Andropogon gerardii*	—	Seed heads in clusters that look like turkey's feet. Leaves turn rust-orange color in fall.	4' to 8'
Black-eyed Susan (W)	*Rudbeckia hirta*	Deep yellow with brown center	Large, showy flowers.	1' to 3'
Butterfly weed (W)	*Asclepias tuberosa*	Orange	Striking flower color. Large pods filled with fluffy seeds.	1' to 2'

(W)-Wildflower, (G)-Grass

Common name	Species	Flower color	Special features	Height
Cardinal flower (W)	Lobelia cardinalis	Red	Ferrari-red flowers attract hummingbirds.	3'
Drooping coneflower (W)	Ratibida pinnata	Yellow with gray center	Large flowers with drooping "petals."	3' to 4'
False dragonhead (W)	Physostegia virginiana	Pink	Large, showy flowers in large clusters.	2' to 4'
Giant blue lobelia (W)	Lobelia siphilitica	Blue	Striking blue flowers.	2'
June grass (G)	Koeleria macrantha	—	Slender, graceful plants. Seed heads are narrow spikes.	2' to 3'
Leadplant (W)	Amorpha canescens	Purple	Small flowers in candelabra-like sprays. Leaves finely divided, gray, hairy.	1½' to 3'
Pale coneflower (W)	Echinacea pallida	Pink	Lovely flower heads with drooping "petals."	2' to 3'
Prairie blazing star (W)	Liatris pycnostachya	Rose-lavender	Flowers produced in long, narrow, cylindrical clusters.	2' to 4'
Purple prairie clover (W)	Petalostemum purpureum	Rose-purple	Flowers produced in dense clusters at tips of plants.	1' to 3'
Rattlesnake master (W)	Eryngium yuccifolium	White	Striking appearance of plant. Pale-colored, yucca-like leaves.	3' to 4'
Rough blazing star (W)	Liatris aspera	Rose-lavender	Long, cylindrical flower clusters. Often used as a cut flower.	1½' to 4'
Switchgrass (G)	Panicum virgatum	—	A large grass. Can be attractive if grouped together. Open seed head.	3' to 6'
White prairie clover (W)	Petalostemum candidum	White	Flowers produced in dense clusters at tips of plants.	1' to 3'
White wild indigo (W)	Baptisia leucantha	White	Long flower spikes point upward above leaves.	2' to 4'
Wild bergamot (W)	Monarda fistulosa	Lavender to pink	Plants form large clumps. Very fragrant foliage.	2' to 3'

A prairie for all seasons
Native prairie plants in the Midwest, cont.

Fall-blooming plants

Common name	Species	Flower color	Special features	Height
Compass plant (W)	*Silphium laciniatum*	Yellow	Large cut leaves often point north-south.	4' to 9'
Culver's root (W)	*Veronicastrum virginicum*	White	Small flowers in large candelabra-like clusters.	3' to 5'
Indiangrass (G)	*Sorghastrum nutans*	—	Large, graceful plants. Bronze fall color of leaves. Seed heads copper and yellow, fluffy.	4' to 8'
Little bluestem (G)	*Schizachyrium scoparium*	—	Leaves turn rust-orange color in fall. Seed heads covered with white hairs, becoming feathery along stems.	3' to 4'
New England aster (W)	*Aster novae-angliae*	Rose-purple	Very showy plant. Fragrant foliage.	1' to 4'
Northern prairie dropseed (G)	*Sporobolus heterolepis*	—	Plants form graceful mounds. Leaves turn rust-orange color in fall. Seed heads fragrant.	2' to 3'
Prairie dock (W)	*Silphium terebinthinaceum*	Yellow	Tall plant. Large leaves at base of plant.	5' to 10'
Rigid goldenrod (W)	*Solidago rigida*	Yellow	Flower clusters flat-topped. Leaves covered with soft, gray hairs.	2' to 4'
Showy goldenrod (W)	*Solidago speciosa*	Yellow	Flower clusters form large, narrow pyramids. Leaves turn red in fall.	2' to 6'

(W)-Wildflower, (G)-Grass

Prepared with Patrick Weicherding, urban and community forestry specialist, UI Cooperative Extension Service; and Kenneth Robertson, botanist, Illinois Natural History Survey.

8 Attract birds and butterflies to your backyard

Attracting birds

"Nature is very much a now-you-see-it, now-you-don't affair," says Annie Dillard in *Pilgrim at Tinker Creek.* "A fish flashes, then dissolves in the water before my eyes like so much salt. Deer apparently ascend bodily into heaven; the brightest oriole fades into leaves."

In your backyard, the same now-you-see-it, now-you-don't principle applies. But there are ways to make the sighting of specific birds an easier, more enjoyable experience. To make your backyard a haven for birds, provide the essentials: food, water, and shelter. Also, birds generally prefer a landscape rich in trees and shrubs.

Trees and shrubs that attract birds

Taller trees

	Food	Cover	Nesting
American beech	Fruit, sap, buds		
Birch, river	Seed	x	
Hackberry	Fruit, sap	x	x
Oak, pin, red, white	Fruit	x	x
Sugar maple	Seed, sap, buds	x	x
White pine (Eastern)	Seed	x	x

Smaller trees

	Food	Cover	Nesting
Alder, smooth, speckled	Seed, buds	x	x
American holly	Fruit	x	x
Cherries (many types)	Fruit	x	x
Crabapples	Fruit	x	x
Eastern red cedar	Fruit	x	x
Flowering dogwood	Fruit, sap		
Eastern hemlock	Seed	x	x
Hawthorns	Fruit	x	x

Shrubs

	Food	Cover	Nesting
Dogwood, alternate-leaf, gray	Fruit, buds	x	x
Elderberry	Fruit	x	x
Honeysuckles	Fruit	x	x
Roses	Fruit	x	x
Serviceberry, downy	Fruit		
Serviceberry, shadblow, smooth	Fruit	x	x
Sumac, shining, smooth, staghorn	Fruit		
Winterberry	Fruit	x	x

These trees and shrubs are the best choices for attracting birds based on the number of bird species attracted and the landscape value of the plant.

SOURCE: Adapted from *Trees, Shrubs, and Vines for Attracting Birds,* Richard M. Degraaf and Gretchin M. Witman, University of Massachusetts Press, 1979.

Bird feeders

There are four basic types of bird feeders:

Gravity feeders. Gravity feeders usually have a roof and either glass or plastic sides so the birds can see the food and you know when it needs to be refilled. These feeders allow for continuous feeding.

Open-shelf feeders. These feeders may or may not have roofs and usually do not have sides except for a small rim, which keeps the seeds from falling or blowing away. Uncovered feeders allow the birds to see danger and are popular because the birds are very visible.

Ground feeders. Simply scattering food over a clear plot of ground constitutes a ground feeder. However, you can also place any platform

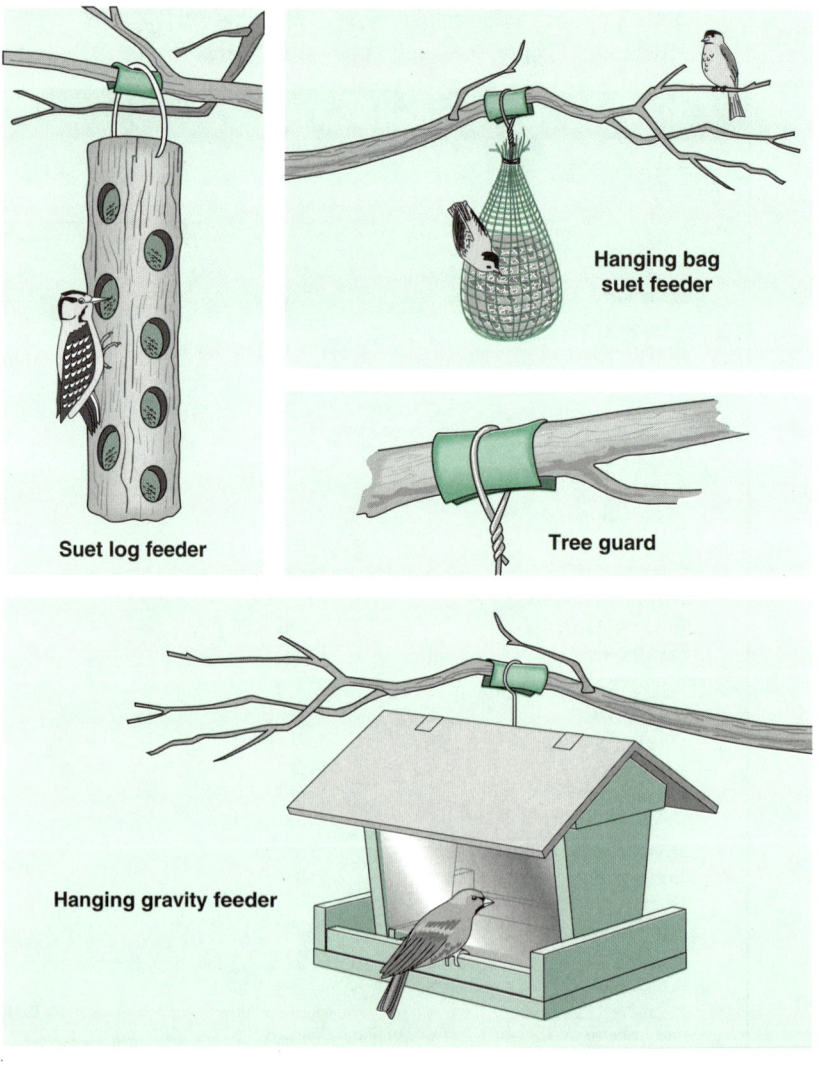

Suet log feeder

Hanging bag suet feeder

Tree guard

Hanging gravity feeder

feeder, with or without a roof, on the ground. One advantage of a ground feeder is that it attracts several species of birds that rarely visit feeders hung from trees, placed on poles, or attached to buildings.

Suet feeders. Suet feeders commonly consist of either a small wire basket or a large mesh bag in which suet is placed. These feeders are either suspended or permanently affixed to the side of a tree, building, or other feeder.

In addition to the four basic types of feeders, there is an endless variety of specialty feeders, such as: pine cones dipped in fat; birdseed logs; strings of peanuts, berries, or other tidbits; or open coconut shells stuffed with an assortment of delectables.

Types of food

The following are the most common food types and the birds they attract.

- Sunflower seeds (unsalted): grosbeaks, cardinals, titmice, chickadees, nuthatches, woodpeckers, finches.
- Millets and small-seed mixtures: cardinals, chickadees, titmice, nuthatches, woodpeckers, native sparrows, finches, juncos, towhees, blackbirds.
- Cracked corn: doves, jays, cardinals, towhees, juncos, native sparrows, woodpeckers, house sparrows, blackbirds, quail.
- Thistle (niger) seed: goldfinches, siskings, redpolls, other finches.
- Shelled peanuts (unsalted): jays, woodpeckers, cardinals, grosbeaks, titmice, chickadees, nuthatches.
- Suet: woodpeckers, titmice, chickadees, nuthatches, starlings, creepers, mockingbirds, wrens, jays.
- Assorted pieces of fruit (including raisins): mockingbirds, waxwings, robins, orioles, starlings.

Other foods that certain species relish include worms, some vegetables, bakery products, cheese, chopped hard-boiled eggs, coconut meat, hominy, peppers, and pumpkin and squash seeds.

A single food type usually does not provide an adequate variety for good nutrition, so develop a blend for the specialized needs of local birds. You might try putting out an experimental feeder with several trays of different foods. Although bakery products may be popular with some birds, they provide little nutritive value. At best, they do a good job of attracting birds to the feeder.

Clean and disinfect feeders regularly to reduce the risk of disease, particularly during summer when mold grows readily on wet bird seed. To avoid this problem, many people feed birds only during the winter.

Also, be sure to provide birds with grit—material used in a bird's gizzard to grind up seeds and other food. Coarse sand is an excellent source of grit. You can mix it directly in with the food or make it available in a separate location. Crushed egg shells also act as a grinding agent and provide a source of needed calcium.

Placement of feeders

Locate feeders about 6 to 12 feet from cover (trees and bushes) to allow feeding birds to escape easily from predators. And don't forget to test the view so you can watch the birds during cold winter weather from your indoor perch. You'll be more motivated to maintain the feeding station if it is within sight.

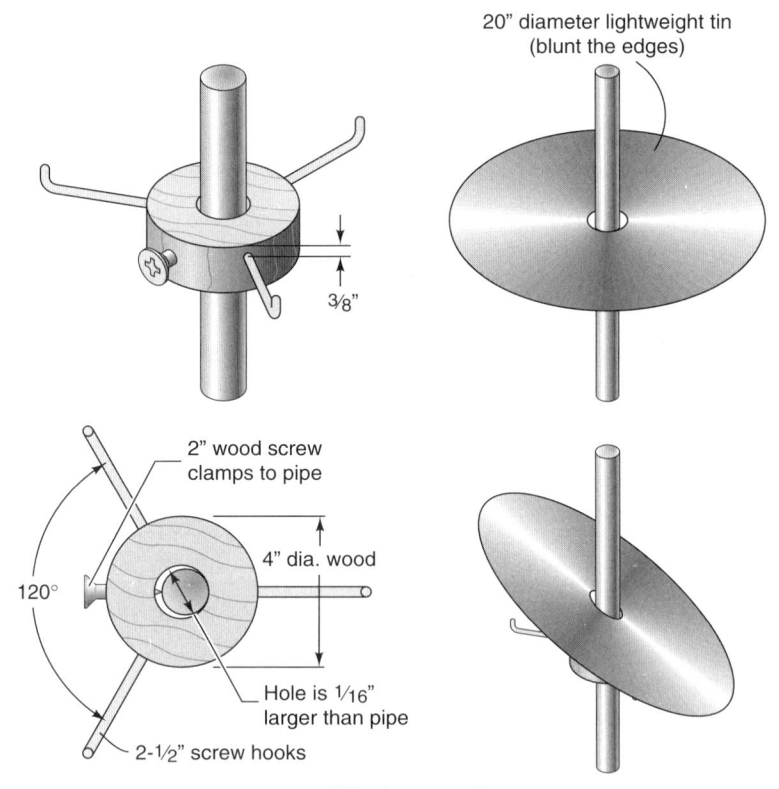

Tipping guard

In the ongoing battle to prevent squirrels from stealing bird food, try installing a tipping guard on the bird feeder pole. Squirrels will have a hard time scrambling over the tipping guard to get at the food. To prevent children and others from getting cut by the sharp tin, be sure to blunt the edges. Using a needle-nose pliers, fold the edges under.

You may want to consider placing different types of feeders in various locations to spread out the population of visitors—cardinals will be attracted to feeders near hedgerows, whereas chickadees are more likely to visit those placed high above the ground in trees.

To prevent squirrels from stealing food from the birds, consider putting a tipping guard on the pole of the bird house (see illustration).

Water

You can fulfill critical water needs with a simple bird bath or ground watering device. Size is not important, but the edges of the bath should slope gradually. Make sure your water source is fresh and clean, especially in the winter when many natural sources are frozen and inaccessible. One way to keep the birdbath from freezing is to use an electric immersion water heater.

Don't forget the butterflies

Although butterflies certainly are quieter, they are no less desirable as backyard visitors than the many birds your careful plans can help to attract. All you need to become a butterfly gardener is a sunny space, good soil, an assortment of nectar-producing flowering plants, and a little bit of hard work. A tolerance for some plants that your garden-club neighbors may label as weeds also doesn't hurt.

You'll want to provide food plants for all stages of a butterfly's life cycle if you want to attract the maximum number of these "flying flowers," as some have called them. The *adult* butterfly can only ingest liquids (nectar), whereas butterfly *larvae* (caterpillars) are more voracious, eating leaves.

In general, plants that bloom for much of the summer and produce large amounts of nectar will attract many adult butterflies. These include butterfly bush, butterfly milkweed, tithonia, and large-flowered zinnias and asters. The following table shows common midwestern butterflies and their preferred food sources.

Remember that butterflies are cold-blooded and need sunlight to warm their flight muscles, so it's best to locate the attractive plants in a sunny area. Also, butterflies enjoy relaxing in the sun when not feeding. Wind and predators can be serious threats to butterflies, so it's a good idea to plant your garden in a protected spot next to a vine-covered fence, a wall, or a windbreak of shrubs or trees.

Food plants for butterflies

Butterfly common name	Season	Larval food	Adult food
Black swallowtail	April to October	Fennel, parsley	Zinnia, aster, tithonia, butterfly weed
Comma	March to November	Hops, nettles, elms	Butterfly bush, sedum, dandelion
Common sulphur	April to November	Alfalfa, clover, vetch, lupine	Goldenrod, dandelion, tithonia, milkweed, phlox
Great spangled fritillary	April to September	Violets	Thistle, vetch, *Monarda*, verbena
Monarch	April to November	Milkweed	Milkweed, butterfly bush, goldenrod, liatris, tithonia, cosmos, mallow
Mourning cloak	Year-round	Willow, elm, poplar, birch, hackberry	Butterfly bush, pussy willow, butterfly weed
Painted lady	April to October	Thistle, burdock, hollyhock	Mint, bee balm, sedum, butterfly bush
Pearl crescent	April to October	Asters	White clover, geranium, butterfly weed, mint
Red admiral	March to November	Nettles, hops	Butterfly bush, milkweed, alfalfa, sedum, mint
Red-spotted purple	April to September	Willow, poplar, cherry	Spirea, lilac, viburnum, privet, butterfly bush
Tiger swallowtail	April to September	Cherry, ash, birch, tuliptree	Butterfly bush, phlox, Joe-Pye weed, bee balm, native honeysuckle, milkweed
Viceroy	May to September	Willow, aspen, poplar, apple, plum	Thistle, aster, Joe-Pye weed, goldenrod, milkweed

SOURCES: *Annotated Checklist of the Butterflies of Illinois*, by Roderick R. Irwin and John C. Downey, Illinois Natural History Survey, 1973; and "Developing a Butterfly Garden," by Michael R. Jeffords and Susan L. Post, *Illinois Steward*, Winter 1994.

Prepared with Philip Nixon, entomologist, UI Cooperative Extension Service, in cooperation with the Division of Natural Heritage, Illinois Department of Conservation.

WHAT ABOUT THE WILDLIFE YOU DON'T WANT?

Laws affecting urban wildlife

A natural backyard habitat for wildlife adds beauty, creates an educational and exciting environment for kids, and provides opportunities for photography and bird watching. However, by improving the habitat for one species of wildlife, there is always the chance that unexpected guests will show up—less desirable or nuisance wildlife. For assistance in dealing with nuisance animals, contact your local animal control department. They can help to remove certain animals in non-lethal ways, and they can also provide information on local, state, and federal laws that protect some species.

Some advice on live traps

Live traps, as the name indicates, preserve the life of the animal. For the most humane treatment of wildlife, follow these guidelines:

- To minimize stress and injury to trapped animals, check live traps at least once and preferably twice a day. When trapping nocturnal animals, check the traps as early in the morning as possible.
- Traps containing animals should be covered with a blanket or placed within burlap sacks. Keeping the animals in darkness and handling the traps gently will help the animals remain calm.
- Relocate captured nuisance animals at least 5 miles from the trap site, and release them in an area where they will not cause a problem for someone else.
- Never release a wild animal that shows signs of sickness, outward aggression, or excessive salivation. It may be rabid and should be taken to a local animal shelter for proper attention and possible disposal.

Common nuisance animals

Bats

Damage/hazards

- A small percentage have rabies. Approach with caution. If rabies is suspected, contact local public health personnel.

Control methods

- Bat-proof building in late fall (after bats leave for hibernation), late winter, or early spring (before bats arrive). Seal openings larger than ⅜-inch to keep bats out. Possible materials: ¼-inch mesh hardware cloth, sheet metal, plywood, aluminum flashing, plastic bird netting.

Comments
- If a bat flies into the house, open doors and windows, and turn off lights. If lights are left on, bat may hide behind drapes or wall hangings.

Chipmunks

Damage/hazards
- Can burrow under buildings and into flower beds and lawns. Can carry disease, so use tongs or gloves when handling chipmunks.

Control methods
- Use live traps.
- Use rat-size snap traps. Cover with a box with 2-inch diameter holes cut into each end to avoid harming nontarget wildlife.

Comments
- Bait traps with nutmeats, sunflower seeds, peanut butter, corn, or rolled oats. Burrows are usually near structures, so smoke bombs are not advised.

Deer mice

Damage/hazards
- Can damage food, furniture.

Control methods
- Use same kind of snap traps or live traps used for house mice.

Comments
- Same size as house mice, but have distinct bicolored tail, body. Bait traps with oatmeal or peanut butter.

Moles

Damage/hazards
- Burrowing moles can disfigure lawns, dislodge plants, and injure plant roots.

Control methods
- Use traps. Harpoon traps are the easiest to use and most available.

Comments
- Poison baits, poison gases, ultrasonic devices, and spinning daisies are all ineffective.

Rabbits

Damage/hazards

- Not a big problem, but occasionally damage plants, trees, gardens.

Control methods

- Remove habitat (heavy vegetation, brush, and piles of debris).
- Use repellents to make trees and shrubs less attractive.
- Place ½-inch mesh hardware cloth around plants. Hardware cloth should be 2 feet high, snug to ground. Rabbits do not burrow.
- Put wire fence around yard or garden.
- Use live traps.

Comments

- Bait live traps with apples, lettuce, carrots.

Raccoons

Damage/hazards

- Nuisance. Raid garbage cans and can beat you to the best sweet corn.

Control methods

- Store garbage in metal or tough plastic containers with tight lids. Wire, weight, or clamp lids down.
- Keep cans on rack or other support to prevent tipping.
- Use live traps.

Comments

- Bait live traps with sweet corn, sardines and other fish, fish-flavored cat food, melon, fried bacon, or cooked fatty meat. Wadded-up foil in trap stimulates their curiosity.

Tree squirrels

Damage/hazards

- Sometimes enter buildings, causing damage. Look for entry points: utility lines, drain pipes, ivy, tree limbs, attic and/or basement vents, and windows in disrepair.

Control methods

- Seal openings with ½-inch hardware cloth.
- If squirrels gain access from trees, prune limbs 6 feet back from house.
- Use rat-size snap traps for red squirrels only.
- Use live traps.

Comments
- Most squirrels are protected animals. Make sure traps do not violate any laws. Check with local game conservation office.

Woodchucks (groundhogs)

Damage/hazards
- Can ruin a garden. Burrow under homes, patios, garages, and stored lumber.

Control methods
- Use live traps.
- Place fumigants (smoke bombs and gas cartridges) in burrows away from buildings. Avoid prolonged breathing of smoke.

Comments
- Fumigants most effective on cool, rainy days.

Prepared with Robert M. Corrigan, animal damage control specialist, Purdue University Cooperative Extension Service.

Kathy Wildman

Life down on the butterfly farm

"If you plant it, they will come," says Kathy Wildman of Sunbury, Ohio.

Wildman should know, for she has been planting an abundance of flowers in her 1-acre native-plant nursery established in the early 1980s, and her flowers have been attracting an equally abundant number of butterflies.

"Imagine a garden where the colors not only wave in the breeze but lift up into the air and float off with the wind," says Wildman, owner of the Hearts and Flowers Butterfly Farm. This is the effect you can get when you plant a butterfly garden.

To start your own butterfly garden, Wildman suggests you find out which butterflies are likely to come to your area, then obtain native flowering plants that will attract those visitors.

"It's important that you plant a mass of color rather than just a dot of color here and there," she says. "This will reinforce the color and help to attract butterflies."

Adult butterflies seek the flowers as a source of nectar, and they provide an important pollination service. But for even greater success in attracting butterflies, Wildman stresses the importance of also providing food for the caterpillars—the larvae that will later transform into butterflies. (See the chart on page 46 for ideas on food sources.)

Wildman recommends that you plant fragrant and colorful flowers that have short tubes and a surface for the butterflies to land on, such as a daisy or buddleia (butterfly bush). As an added attraction for butterflies, provide a shallow water source or mud puddle because butterflies (especially males) often congregate at puddle-side "drinking clubs."

Butterfly houses are sold commercially as a site for overwintering. But according to Wildman, these houses are often promoted with little knowledge about butterfly biology.

She says that when butterflies overwinter they look for protected crevices, often against a building or at the edge of a woodlot. However, people make the common mistake of placing their butterfly home in the open sun—usually on a post in the middle of the flower garden.

In the Midwest, the houses have limited use because few butterflies overwinter as adults in the region; most migrate from the south in the spring. For butterflies that do overwinter in the Midwest, an equally effective overwintering site is a brush or log pile.

How successful you are in attracting butterflies depends on your location. If you live in an industrial zone, it may take longer for butterflies to find you. But if you live in an area where the neighbors also plant flowers, Wildman says you may see a healthy population of butterflies.

How long the butterflies stay in your yard depends on the species. It also depends on whether you supply them with more than just nectar-producing flowers. These delicate creatures have other needs as well—food for the larval stage and shelter for all stages of their life. Because insecticides can kill butterflies, Wildman recommends that you use them only as a last resort.

A healthy backyard habitat will also attract spiders, birds, and beneficial insects, which will help keep pests under control, she adds. What's more, by learning to appreciate one of the most popular and attractive of insects—the butterfly—Wildman says you may learn to value all insects.

"Who knows?" she says. "Maybe you'll learn to love spiders next."

Using alternatives to traditional pesticides and fertilizers—where practical

Allowing nature to run its course doesn't always work out as nicely as we might hope, because nature is rarely neat and tidy. Most people discover that some form of pest control may be desirable or even essential around the home. And if you want an attractive lawn or a productive garden, fertilizers may be needed.

Keep in mind, however, that it is always good to choose pest-control and fertilizer options that have the fewest negative impacts around your home, yard, and garden. The following section offers a few of the most realistic ideas.

9 Scout for pests on the landscape

Looking for trouble

In the world of yards and gardens, the best way to prevent trouble is to go out looking for it. Regularly "scout" your lawn and garden for signs of damage or the culprits themselves—weeds and insects.

In addition to helping you catch problems in the early stages, scouting is one of the best ways to cut down on the use of pesticides, for a significant percentage of lawn and garden chemicals are put down as precautionary measures—before a problem even shows up. By scouting, you can wait to make sure a problem exists before you try to fight it. And even then, there may be ways to combat the problem without resorting to pesticides. Consider the alternatives described in Chapters 10–15.

Scouting also makes it possible to "spot-treat" only those areas that require attention, rather than broadcasting chemicals across an entire lawn or garden.

How much can you tolerate?

Spotting signs of damage doesn't automatically mean you should take action. It all depends on the type of pest, the type of damage caused, the severity of infestation, and your own personal preferences. Determine whether the damage falls into any of the following categories:

Economic damage. Some insects, such as termites, cause economic damage to a home, yard, or garden.

Medical risks. Some insects and related creatures are a concern because they carry disease. A good example is the tick, one species of which is responsible for transmitting Lyme disease.

Aesthetic damage and nuisance problems. Some homeowners do not want to see holes eaten into leaves or all of the leaves eaten off of a plant—particularly the leaves of ornamental plants—even though the plant may be able to tolerate the damage. Even normally desirable plants may become a nuisance if they begin to take over an entire yard.

After you identify a problem, determine the potential for damage and how much you or your landscape can tolerate. Once you know what you're up against, find out if the damage can be controlled with the less toxic strategies described in this book. Use pesticides as a last resort.

For assistance in scouting, identifying pests, and assessing potential damage, contact your nearest Cooperative Extension Service office. Numerous scouting guides are also available to help you identify pests and the damage they cause.

> ✓ **CHECK IT OUT**
> For ideas on toxicity levels of insecticides: pages 92, 97.
> For help in identifying indoor pests: page 235.

Prepared with Philip Nixon, John Lloyd, and Rick Weinzierl, entomologists, UI Cooperative Extension Service.

Additional source: Common-Sense Pest Control, *by William Olkowski, Sheila Daar, and Helga Olkowski, The Taunton Press, 1991.*

10 Consider microbial insecticides

Pros and cons

Microbial insecticides battle damaging insects by enlisting the aid of microscopic living organisms—viruses, bacteria, fungi, protozoa, or nematodes. They are "unconventional" insecticides, but they can be applied in conventional ways—as sprays, dusts, or granules.

Microbial insecticides have several advantages and disadvantages when compared to typical, synthetic insecticides.

Advantages

Safety to humans. The organisms used in microbial insecticides are essentially nontoxic and nonpathogenic (do not pose risk of disease) to wildlife, humans, and other organisms not closely related to the target insect.

Selectivity. Most microbial insecticides are toxic to a single species or group of insects. This means you can often target a pest without the risk of killing beneficial insects in the process.

Compatibility with conventional pesticides. Most microbial insecticides can be used in conjunction with conventional insecticides. The residues of conventional insecticides usually do not damage or deactivate microbial insecticides.

Impact on harvest. Because microbial insecticide residues present no hazards to humans or animals, microbial insecticides can be applied when a fruit or vegetable is almost ready for harvest.

Longevity. In a few cases, the microorganisms used in these products can become established in an insect population or its habitat. This means the insecticide can provide control for several weeks or seasons.

Disadvantages

Selectivity. Although there are advantages to a microbial insecticide's ability to target a single species or group of insects, there are also disadvantages. If several types of insects are causing damage to your lawn or garden, a single microbial insecticide will not be able to handle them all. Conventional insecticides can be limited in this way too, but the negative side of selectivity is more noticeable for microbials.

Rapid breakdown. Heat, drying out, or exposure to sunlight reduces the effectiveness of several types of microbial insecticides. Therefore, proper timing and application are especially important for some products.

Availability and cost. Because several microbial insecticides are species-specific, their potential market is limited. Consequently, some products are not widely available or are relatively expensive.

Bacteria

The most popular microbial insecticides in the United States are preparations of the bacterium *Bacillus thuringiensis,* better known as *Bt.* Bacterial insecticides must be eaten by the pest for them to do their job. They do not poison the pests on contact.

Of the various *Bt* products, formulations from *Bacillus thuringiensis* var. *kurstaki* are the most widely used. These products are sold under such trade names as Dipel, Biobit, Javelin, Full-Bac, Thuricide, Worm Attack, and Caterpillar Killer. They are pathogenic and toxic only to caterpillars—the larvae of butterflies and moths. Other varieties of *Bt* are available for the control of Colorado potato beetle larvae and certain mosquito larvae.

The following are some key application tips:

- Because each type of *Bt* insecticide controls only certain insects, you must identify the target pest and confirm that the product label says it will control that insect.
- *Bt* products work on the larval stage of a pest, but most do not kill adult insects.
- *Bt* products are applied on the plant surface, so they typically are not effective on pests that bore into the plant tissue without first eating some foliage.
- Thoroughly cover all plant surfaces with *Bt,* even the undersides.
- Treat in the late afternoon or evening, because sunlight deactivates the chemical. Treating on cloudy (but not rainy) days provides a similar result. Also, encapsulated products give the chemical more protection from the sun's ultraviolet radiation.
- Although microbial insecticides are safer than conventional insecticides, bacterial spores, mold spores, and virus particles can cause allergic reactions when inhaled or rubbed on the skin. Wear gloves, long sleeves, and long trousers during application, and wash thoroughly after completing the application. When appropriate, wear a dust mask or respirator to keep from inhaling spray mists or dusts.

Viruses

Viruses, like bacteria, must be ingested by the insect. They often cause dramatic natural disease outbreaks among insect populations. But unlike bacterial insecticides, the development of virus-based insecticides has been limited.

Some important pests for which viral insecticides have been developed include the gypsy moth, pine sawflies, and the codling moth. If these products become more widely available, their effective use will depend on the applicator remembering the following key facts:

- Most viruses are effective only against the immature stages of specific insects.
- You must match the virus and the target species correctly.
- Virus particles are killed by the sun's ultraviolet radiation. Therefore, treat in the evening or on cloudy days.

Fungi

Like viruses, fungi create natural epidemics among insect populations, often killing a high percentage of the population. Fungal pathogens have the greatest impact on insect populations during cool and humid weather. Only a few fungal insecticides are currently available in the United States (as of 1995). *Metarhizium anisopliae,* a fungus, is the active ingredient in the Bio-Path cockroach control chamber, and *Beauveria bassiana* is under development for registration and commercial use.

Protozoa

Protozoan pathogens infect a wide range of insects. Some of them can kill insects rapidly, but most of them are known for their chronic, debilitating effects. Consequences of protozoan infections include a shorter life span, less feeding, and a reduction in the number of offspring.

Nosema locustae, a protozoan pathogen of grasshoppers, is sold under the trade names NOLO bait and Grasshopper Attack for rangeland and backyard grasshopper control. Grasshoppers die several weeks after initial infection; this delay and the nature of grasshopper behavior reduce the value of these products for grasshopper control in lawns and gardens.

Nematodes

Technically, nematodes are not microbial agents—they are multicellular roundworms. But being nearly microscopic in size, they are used much like true microbial insecticides.

The nematode species most commonly used to control pests are *Steinernema carpocapsae* and *Heterorhabditis heliothidis.* In general, these nematodes infect a wide range of insects. On a worldwide basis, laboratory or field applications have been effective against more than four hundred insect species, including numerous beetles, fly larvae, and caterpillars.

Large-scale production of nematodes has begun, and a few insecticides containing nematodes have been marketed. But their effectiveness is limited by temperature and moisture conditions. They are generally effective only when used to control insects in moist soil or other protected habitats. Nematodes will survive for up to several months in cool, moist soils, but homeowners should probably expect that nematodes will persist for only two to four weeks after applications to control soil insects.

Research is currently targeting the use of nematodes for the control of annual white grubs in irrigated lawns and grass sod. If nematodes are applied when grubs are present, they would need to remain active only for several days to reduce grub damage. (Homeowners and turf managers can use irrigation to keep soils moist enough for the nematodes to remain active.) Unfortunately, current studies indicate that white grubs in lawns do not appear to be very susceptible to control by nematodes.

On the other hand, nematodes have been used effectively in trials against root weevil larvae and similar soil insects that attack nursery plants, ornamental plantings, garden crops, and potted plants. The success of these treatments depends primarily on the ability to maintain soil moisture wherever nematodes are applied.

Nematodes are not appropriate for termite control. And until further research provides data on specific pests, the use of nematodes to control most above-ground insects is not recommended.

Nematode products face little regulation, so there is a risk of inferior products and exaggerated claims. These products offer a great deal of potential, but their effectiveness is not as unlimited as some promotions suggest.

For more information

For more detailed discussions of these products, the North Central Region of the Cooperative Extension Service offers *Alternatives in Insect Management,* a 73-page manual. To order copies, call (217)333-2007.

Microbial insecticides

Bacteria

Product	Pests controlled
Bt (Bacillus thuringiensis var. kurstaki)*	Common caterpillar pests, such as: Bagworms Cabbage looper Diamondback moth European corn borer in corn Fall webworm Gypsy moth Imported cabbage worm Mimosa webworm Spring and fall cankerworm Spruce budworm Tent caterpillars Tomato fruitworm
Bacillus thuringiensis var. san diego or tenebrionis	Colorado potato beetle larvae Elm leaf beetle adults and larvae
Bacillus popilliae and Bacillus lentimorbus	Larvae (grubs) of Japanese beetle

Nematodes

Steinernema carpocapsae and other Steinernema species	Larvae of a wide variety of soil-dwelling insects.

Protozoa

Nosema locustae	Grasshoppers Mormon crickets

Viruses

Codling moth granulosis virus (GV)	Codling moth caterpillars
Gypsy moth nuclear polhedrosis virus (NPV)	Gypsy moth caterpillars
Pine sawfly NPV	Pine sawfly larvae

*Common caterpillars that Bt does not control in typical usage include: corn earworm (on corn), codling moth, peach tree borer, and squash vine borer.

Prepared with Rick Weinzierl, entomologist, UI Cooperative Extension Service.

Niles Kinerk

Organic pesticides: Patience pays off

"It takes steady nerves to use organic pesticides," says Niles Kinerk, president of Gardens Alive! in Lawrenceville, Indiana.

When using a microbial pesticide such as *Bt*, you may not see anything happen for twenty-four hours after spraying the insects, Kinerk says. But there is no need to panic—the pest larvae should start to turn dark and then die between twenty-four and forty-eight hours.

"In contrast, a synthetic pesticide such as Sevin gives quick results," he notes. "You spray one row, and by the time you come back down the next row, the beetles you sprayed will be lying on their backs with their legs in the air."

But Kinerk says it is more than worth it to trade the fast action of synthetic pesticides for alternative pesticides, which are generally less toxic. And he says an increasing number of people agree. They are deciding that traditional pesticides may be "an unnecessary risk" in the home, yard, and garden.

Kinerk started producing and distributing alternative pest-control products in 1984, because he says organic products were not easy to obtain at the time. Today, Gardens Alive! includes 20 acres where products are tested. The Gardens Alive! insectary also rears nematodes—tiny roundworms that are shipped on a sponge. After the sponge is soaked in tepid water for five to ten minutes, the nematodes migrate out. You then mix the nematodes with water and apply them in a sprayer or sprinkling can to combat soil insects.

Kinerk considers the "selectivity" of many alternative pesticides an important advantage. Selectivity means that the pesticide kills certain pests, rather than a wide range of insects. For instance, the most common strain of the microbial insecticide *Bt* kills only the larval (caterpillar) stage of butterflies and moths.

"Many lady beetles and other beneficial insects would be dead if you used traditional insecticides," he says. "Without natural predators such as lady beetles to help keep pests in check, you couldn't spray enough to control pests."

11 Consider botanical insecticides and insecticidal soaps

What are botanicals and insecticidal soaps?

Botanical insecticides, sometimes referred to as "botanicals," are naturally occurring insecticides derived from plants. Insecticidal soaps are soaps that have been formulated specifically for their ability to control insects. These products are useful alternatives in insect management, but you should judge them according to their traits—not simply by their "natural" source.

Advantages

Rapid degradation. Botanical insecticides and insecticidal soaps degrade readily in the sunlight, air, and moisture. This means they break down rapidly into less toxic or nontoxic compounds, posing less risk to nontarget organisms. Soaps and many botanicals may be applied to food crops shortly before harvest without leaving excessive residues.

Rapid action. Botanicals and soaps may not kill an insect for hours or days, but they act very quickly to stop its feeding.

Low toxicity to mammals. Most botanicals and insecticidal soaps are low to moderate in toxicity to mammals. But there are exceptions, so check the label and handle them with due caution.

Low toxicity to plants. Most botanicals do not damage plants. Insecticidal soaps and nicotine sulfate, however, may be toxic to some shrubs, houseplants, and other ornamental plants.

Disadvantages

Rapid degradation. Rapid breakdown of botanicals, though a benefit in some ways, requires you to be more precise in your timing of applications. You also may need to make more frequent applications if you observe additional damage.

Toxicity. Botanicals are generally less toxic to humans than typical synthetic pesticides, but there are exceptions. For example, nicotine in its pure form is considered a class I (most dangerous) poison. Both inhalation and dermal (skin) exposure to nicotine preparations can cause death. Also, rotenone is similar in toxicity to the common synthetic insecticides carbaryl (Sevin) and diazinon.

Cost and availability. Botanicals tend to be more expensive than synthetic pesticides, and some are not as widely available. The potency of some botanicals may differ from one source or batch to the next.

Lack of test data. Data on effectiveness and long-term toxicity are unavailable for certain botanicals. Also, tolerance levels for pesticide residues on food crops have not been established for some botanicals.

State registration. Several botanicals are registered by the U.S. Environmental Protection Agency (EPA) and are available by mail order, but they are not registered for legal sale in some states.

Botanical insecticides

From 1945 to the 1970s, the only botanicals in widespread use were pyrethrins and nicotine. Since the 1970s, however, interest in botanicals has increased, primarily because of environmental concerns. Following are the most common botanical insecticides in use.

Pyrethrum and pyrethrins. Pyrethrins are components of pyrethrum—a dust or extract derived from the pyrethrum daisy. Pyrethrum products cause an immediate "knockdown" paralysis of insects, and most of them contain the synergist piperonyl butoxide (PBO) to prevent insects from recovering. (Synergists are compounds that improve the action of insecticides, usually by slowing the process in which an insect detoxifies or breaks down poison.)

Pyrethrins with PBO are registered for use on pets and livestock to control fleas, flies, and mosquitoes. They are also registered as indoor household sprays, aerosols, and "bombs" for the control of various flying insects, fleas, and (less effectively) ants and roaches. Pyrethrins are sometimes combined with rotenone and ryania or copper for general use in gardens.

Rotenone. Rotenone is derived from the roots of the tropical legumes *Derris* and *Lonchocarpus*. In insects, rotenone works primarily on nerve and muscle cells and causes poisoned insects to stop feeding. Insects die several hours to a few days after exposure. Rotenone is particularly effective against leaf-feeding beetles and certain caterpillar pests. It is also very toxic to fish.

Several products combine rotenone with pyrethrins, ryania, copper, or sulfur for general insect and disease control in gardens and orchards. Rotenone persists for a few days on treated foliage—longer than most other botanical insecticides.

Sabadilla. Sabadilla, a compound found in the ripe seeds of a tropical lily, causes loss of nerve function and paralysis in insects. Some insects die instantly; others are paralyzed several days before dying.

Sabadilla is effective against certain "true bug" insects such as harlequin bugs and squash bugs, which are difficult to control with most other insecticides. It is registered by the U.S. EPA for use on crops, including squash, beans, cucumbers, melons, potatoes, turnips, mustard, collards, cabbage, broccoli, citrus, and peanuts.

Sabadilla is highly toxic to honey bees, so avoid spraying it when bees are present. Also, sabadilla can severely irritate skin and mucous membranes, and it is a powerful sneeze inducer, so avoid contact with sprays or dusts.

Ryania. Ryania, taken from the woody stems of a South American shrub, causes insects to stop feeding soon after ingestion, but it does not cause paralysis. It may be used on citrus, corn, walnuts, apples, and pears for the control of such pests as citrus thrips, European corn borer, and codling moth. Like rotenone, ryania degrades more slowly than many botanicals.

Nicotine. Nicotine, a well-known component of tobacco, is an extremely fast-acting nerve toxin in both insects and mammals. It is used in greenhouses and gardens as a fumigant and contact poison to control soft-bodied sucking pests such as aphids, thrips, and mites. Although home gardeners sometimes prepare "tobacco teas" to control garden or houseplant pests, this practice is not recommended. These teas are not as toxic as nicotine sulfate sprays, but any nicotine solution that is toxic enough to kill insects is also toxic enough to harm humans.

Citrus oil components: limonene and linalool. Limonene and linalool are especially common in flea dips and pet shampoos. Their acute toxicity to mammals is low, but they should not be considered nontoxic because some questions about chronic exposure remain unanswered. Studies indicate that crude citrus oil formulations (containing other compounds in addition to limonene and linalool) are more toxic and may pose much greater risk.

Other essential plant oils: "herbal" repellents and insecticides. The most common essential oils used as insect repellents are the oils of cedar, lavender, eucalyptus, pennyroyal, and citronella. They are used mostly on or around pets and humans to repel fleas and mosquitoes.

With the exception of pennyroyal, essential oils are thought to pose little risk to people or pets, but they should not be used above recommended rates. Oil of pennyroyal contains a potent toxin that can cause death in humans at doses as low as 1 tablespoon when ingested. Although pennyroyal's toxicity is low when it is absorbed through the skin, some cats are susceptible to poisoning by topical application, possibly because they ingest it during grooming.

Neem. The compound azadirachtin, derived from the leaves, bark, and seeds of the neem tree, is becoming more widely registered and available. It is very low in toxicity to humans and has been used medicinally in some countries. Sold under various names, including Align, Azatin, Neemix, and Bio-Neem, it may be used according to label directions on several food crops and ornamental plants to control whiteflies, thrips, mealybugs, and other pests.

Insecticidal soaps

Despite many years of use, the manner in which insecticidal soaps work still remains somewhat unclear. Some evidence indicates that soaps enter the insect's respiratory system and cause internal cell damage.

The toxicity of insecticidal soaps to humans and other mammals is basically the same as that of any soap or detergent. Ingestion causes vomiting and general gastric upset but appears to have no serious systemic consequences.

Insecticidal soaps are used to control soft-bodied pests such as aphids, thrips, scales (crawler stage only), whiteflies, leafhopper nymphs, and mites. Soaps are effective only against those insects that come into direct contact with sprays before they dry.

Soaps are not very effective against most insects with harder bodies, such as adult beetles, bees, wasps, flies, or grasshoppers. (They can be effective against adult boxelder bugs, however.) In addition, highly mobile insects may escape soap spray applications by flying away. Soaps generally do not harm adult beneficial insects, but they can kill the immature, soft-bodied forms of beneficial species, just as they kill pests.

Soaps are particularly useful for controlling insects that damage ornamental plants and houseplants, though they can be toxic to some plant species. Plants that have hairy leaves may be more susceptible to soap injury than smooth-leaved plants.

Pests controlled with insecticidal soaps

Effective against soft-bodied pests, such as:

- Aphids
- Leafhopper nymphs
- Mites
- Scales (crawler stage only)
- Thrips
- Whiteflies

For more information

For more detailed discussions of these products, the North Central Region of the Cooperative Extension Service offers *Alternatives in Insect Management,* a 73-page manual. To order copies, call (217)333-2007.

Prepared with Rick Weinzierl, entomologist, UI Cooperative Extension Service.

DON'T PLAY CHEMIST

Mixing household products can be deadly

If you are tempted to create your own pesticides or household cleansers by mixing household chemicals—products such as cleaning agents, fuel oils, polishes, powders, or solvents—think again.

Creating poisons for pest control or homemade cleansers can result in substances that are toxic, possibly deadly, to humans and pets. Such chemical combinations also can pose serious environmental risks.

Most homeowners know, for example, that mixing chlorine bleach with ammonia can produce deadly chlorine gas. Other chemical interactions may be less well known—but equally dangerous. Even household chemicals that are generally not considered to be highly toxic in normal uses—many cleaning agents, for example—pose much greater risk if they are used as homemade insecticides to treat food products. Ingesting residues of these homemade insecticides can be very dangerous. In fact, homemade insecticides can be far more dangerous than commercial pesticides, which have known toxicity levels as a result of testing.

'Natural' does not mean 'nontoxic'

Remember, too, that creating homemade pesticides with natural compounds may not be any safer—and may even be more dangerous—than using commercial pesticides. Natural compounds are not necessarily less toxic to humans than synthetic ones. Some of the most deadly, fast-acting toxins and some potent carcinogens (cancer-causing agents) occur naturally. Botulism toxins and aflatoxins (from microorganisms) and strychnine (from plants) are examples of natural toxins.

Some homeowners have been known to soak cigarettes or cigars (made from naturally grown tobacco) in water to produce a homemade, nicotine-based insecticide. But nicotine is far more toxic, especially at relatively high concentrations, than most people realize. Nicotine sprays that are strong enough to kill insects can also be deadly to humans and pets, even as a result of spills on the skin.

Although it may seem tempting to trust homemade remedies over "chemical insecticides," playing chemist with household substances rarely yields a safe and effective insecticide. The bottom line: do not assume that any recipe that combines household products to make pesticides or cleaners is safe.

Prepared with Rick Weinzierl, entomologist, UI Cooperative Extension Service.

12 Conserve beneficial insects

Predators and parasites

Some bugs are on the side of homeowners. They are known as beneficial insects, or natural enemies of damaging insects, and they fall into two main categories: predators and parasites.

Predators hunt and feed on other insects. Common examples include praying mantids, lady beetles, and green lacewings.

Parasites hatch from eggs inside or on another insect, and they eat their host insect as they grow. Many parasites are tiny wasps that don't sting humans but lay their eggs inside other insects.

Releasing beneficial insects into your yard or garden may have some benefit, but your best bet is to *conserve* the beneficial insects already there. Keeping these beneficial insects alive may help to keep pest problems at an acceptable level so you can reduce insecticide use.

There are no easy answers for keeping beneficial insects alive, but the following guidelines should be a good start.

Five steps to conserving beneficial insects

1. *Recognize the difference between pests and beneficial insects.* To determine whether certain insects will cause problems in your yard or garden, you must be able to distinguish a pest from a beneficial insect. For assistance and information on identifying beneficial insects, check for helpful references at your local library and bookstore, or contact your nearest Cooperative Extension Service office.

2. *Minimize insecticide applications.* Many insecticides are nonspecific, which means they kill all insects, including the ones you want to keep. Also, many beneficial insects take longer to return to an area treated with insecticides than do insects that feed on plants. Therefore, treat yards and gardens only when pest populations are great enough to cause significant damage. In addition, use the most accurate application methods you can.

3. *Use selective insecticides, and apply them selectively.* The ideal insecticide would be one that kills only the particular insect you have targeted. Although most insecticides are not this selective, shop carefully to find one that is more specifically aimed at your pest problem (such as *Bt* products for caterpillar control). Insecticides that must be applied directly to the insect or that break down quickly on treated surfaces (such as natural pyrethrins and insecticidal soaps) also kill fewer beneficial insects.

Beneficial insects

Big-eyed bug

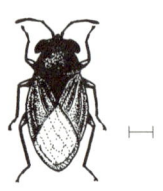

Common damsel bug

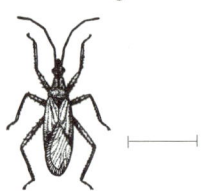

Chinese praying mantid

Egg case with newly hatched nymphs

Adult
Actual size: about 3 to 4 inches

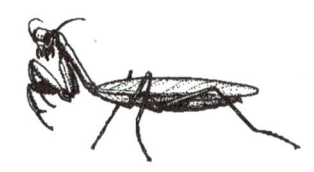

Common green lacewing

Egg

Larva ("aphid lion")

Adult

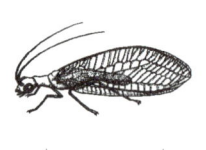

Convergent lady beetle

Larva

Pupa

Adult

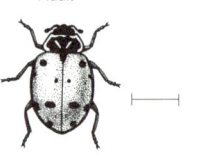

Ground beetle

Minute pirate bug
Actual size: the size of a dash

⊢———⊣ Indicates actual size of adult insect

Predatory mite
Adult and egg
Actual size: the size of a period

Rove beetle

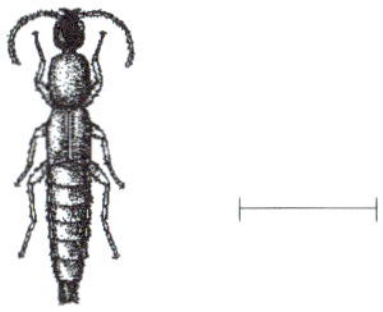

Some important lady beetles

Spotted lady beetle

Twospotted lady beetle

Seven spotted lady beetle

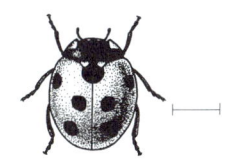

Twice-stabbed lady beetle

Spined soldier bug, feeding on a Mexican bean beetle pupa
Adult

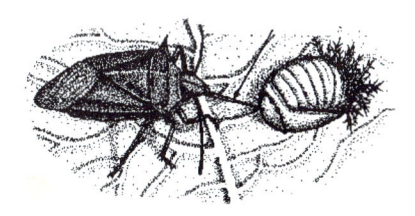

Syrphid fly
Larva *Adult*

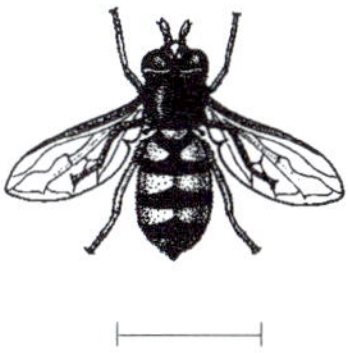

Trichogramma wasp
Adult
Actual size: the size of a period

⊢——⊣ Indicates actual size of adult insect

> ✓ **CHECK IT OUT**
> For details on *Bt* products, pyrethrins, and insecticidal soaps: pages 57, 63, 65.

4. *Maintain the habitat of beneficial insects.* Beneficial insects are often slow to colonize a yard. The best way to make sure they are nearby is to maintain their natural habitats. You can do this by maintaining a diversity of plantings. The greater the diversity, the greater the chance that you will meet the needs of beneficial insects—food, water, and shelter.

5. *Provide pollen, nectar sources, or artificial food.* Adults of some beneficial insects need to feed on pollen and nectar. Plants with very small flowers (such as some clovers and Queen Anne's lace) and some flowering weeds in and around the yard may help preserve a diversity of insect life.

   In addition, artificial food supplements that contain yeast, whey proteins, and sugars can attract or increase the numbers of adult lacewings, lady beetles, and syrphid flies—all beneficial insects. These food supplements can be purchased from suppliers who serve "organic" farmers and gardeners.

Important considerations

Although the first three points mentioned above are excellent ideas for any homeowner, the last two must be thought through carefully. Maintaining habitats and providing artificial food do more than attract just beneficial insects. These strategies can attract pests as well.

Also, keep in mind that although it is important to bring pest populations below damaging levels, it is not necessary to completely eradicate undesirable insects. In fact, leaving some pests alive will help maintain the populations of beneficial insects. If you eradicate all pests, then beneficial insects will leave in search of other food sources.

Limitations

In many cases, conserving beneficial insects will provide natural control of insect pests. But natural control does have its limitations.

Predators and parasites work slowly. With insecticides, homeowners are accustomed to seeing quick results. In contrast, it may take weeks or even seasons for beneficial insect populations to build up.

When pests become few and far between, beneficial insects leave the area in search of more prey. So you are always left with at least a few undesirable insects still in the yard.

Despite the limitations, maintaining or cultivating populations of beneficial insects is not difficult. And it could reduce your exposure to insecticides.

Releasing beneficial insects

If you are interested in releasing beneficial insects in your yard, you can purchase them from a commercial insectary (insect-rearing facility). They are usually shipped to you in an inactive stage—eggs, pupae, or chilled adults—ready for release.

One approach is to release low numbers of beneficial insects one or two times in a season; the goal is for the beneficial insects to establish themselves and suppress undesirable insects for the entire season—or longer. Another approach is to release large numbers of beneficial insects to overwhelm and rapidly reduce the pest populations. With large releases, beneficial insects may or may not establish themselves for the entire season, and costs are much higher.

For readers interested in obtaining a list of beneficial insect suppliers, the California Department of Food and Agriculture offers a helpful booklet, *Suppliers of Beneficial Organisms in North America.* Single copies are available free of charge from Biological Control Services Program, 3288 Meadowview Road, Sacramento, CA 95832.

Prepared with Rick Weinzierl, entomologist, UI Cooperative Extension Service.

DO COMPANION PLANTS DETER PESTS?

"Companion planting" is the practice of growing specific vegetables, flowers, and herbs next to one another for their possible mutual benefit. Companion plantings are intended to improve the growth of particular crops, or in some cases they are expected to benefit crops by releasing a substance that deters an insect or inhibits a disease.

For example,

- Many gardeners plant marigolds not only for their attractive flowers but also to repel harmful insects.

- Interplanting clovers or other legumes with corn provides nitrogen and can reduce damage to corn by European corn borer.

- Clovers and small-flowered plants in the carrot family provide nectar and pollen for beneficial insects that help to control other insects.

As promising as these examples might sound, the story on companion plants is not quite so simple. Although many plants do contain or emit chemicals that are detrimental to some pests, the benefits of common companion plants are often less than gardeners might expect.

For instance, certain marigolds contain high concentrations of substances that kill or repel aphids and nematodes, but you won't see these benefits by using just a few plants of most commercial varieties. Common varieties of marigolds do not contain high concentrations of the natural pesticidal compounds; in fact, their flowers attract moths that may lay eggs on nearby plants. Even the "right" variety of marigolds may provide nematode control only when pure stands are grown for a season or more.

Perhaps it is best to choose companion crops on the basis of their more obvious impacts. Where resources are plentiful—as is usually the case in most midwestern yards and gardens—companion plantings can add diversity and beauty.

Prepared with Rick Weinzierl, entomologist, and James Schmidt, horticulturist, UI Cooperative Extension Service.

13 Monitor and control insects with traps

Chemical communication

Insects use chemicals to transmit signals to each other for a variety of purposes: sending out alarms, attracting a mate, marking trails, and tracking down food, to name just a few. By tapping into this chemical communication network, you can monitor or sometimes reduce insect problems.

One way to do this is by luring insects to traps baited with synthetic versions of the chemicals that they use to communicate with each other. The chemicals most often used for this purpose are "pheromones"—chemical signals produced and received by members of the same insect species. Traps are baited with two types of synthetic pheromones: sex pheromones (chemicals used to attract mates) and aggregation pheromones (chemicals used to attract more insects to a food or nesting site).

Using baited traps to monitor pests

The most common use of pheromones in insect management is for *monitoring* insect populations. Traps, usually made from paper or plastic, capture insects that are lured in by chemical attractants. Many traps are coated with an adhesive to capture insects, whereas other traps allow an insect to enter through a narrow, funnel-shaped opening into an enclosure from which they cannot escape.

You can use traps to detect a particular insect in your yard or garden and then determine if and when you need to apply an insecticide. For homeowners and gardeners, pheromone-baited traps are particularly useful in determining when to control such insects as the codling moth in apples, and the peachtree borer in peaches and related fruit and ornamental trees.

Using baited traps for mass trapping

Depending on the pest, placing numerous traps throughout the yard or garden can sometimes remove enough insects to limit their damage. This is called "trapping out," or "mass trapping."

If you use traps baited with an aggregation pheromone, you can reduce insect populations by capturing male and female insects before they eat much of a crop or reproduce. If you use traps baited with a sex pheromone, the key is capturing male insects before they mate with the females.

For mass trapping to adequately reduce pest populations, however, you need to use a sufficient number of traps that are very efficient at capturing a high volume of insects. Mass trapping is most likely to succeed when: (1) the population density of the insect you are trying to control is low; and (2) insect migration into the yard does not replace insects as fast as you can trap them.

Pheromone-baited traps are sometimes used for Japanese beetle control, but their effectiveness in actually reducing the populations of this pest is questionable.

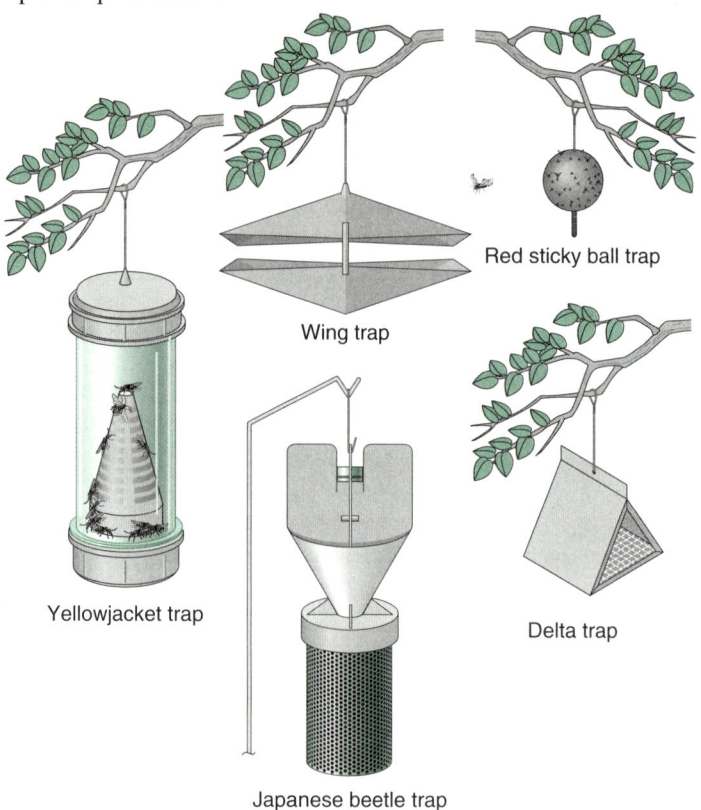

Traps can be used to either monitor insect infestations or to control insect populations.

Using attractants in poison baits

Combining insect attractants with poison (insecticides) has been practiced for centuries. Poisoned baits can sometimes be used effectively at low rates, often without leaving toxic residues on plants or animals. In some cases, insects take baits back to their nest, poisoning other insects.

You can purchase insecticidal baits to control several insects, including the house fly, slugs, certain ants, cockroaches, and yellowjackets.

Using colored objects to attract insects

Specific colors are attractive to some day-flying insects. For example, yellow objects attract many insects and are often used in traps designed to detect the flights of aphids and to control adult whiteflies in greenhouses. Red spheres, coated with an adhesive and baited with an attractant, can be used to capture apple maggot flies and limit damage to fruit. For trapping to prevent apple maggot damage, traps must be in place before flies begin to lay eggs on fruit (by mid-June to early July). Use one to three traps per tree or one every 15 feet in the trees along the edge of an orchard.

Using lights to attract insects

One fairly common strategy is to use lights to lure insects into a device that electrocutes them—"bug zappers." But these devices are ineffective in most instances, as explained on page 76. Nevertheless, the positioning of outdoor lights can be important. Placing outdoor lights several feet away from doors reduces the entry of insects into homes and apartments when insects are active around the lights. In addition, yellow light bulbs attract fewer insects than white incandescent lights or fluorescent bulbs.

Fly traps

House flies commonly land and rest on narrow, vertical objects. That's why hanging sticky "fly strips" is sometimes effective in small, closed areas where fly populations are low—a closed porch or similar indoor area.

Flies often land near other flies, so strips that have captured a few flies and strips that bear pictures of flies may be more effective than clean strips. Hang strips so people do not inadvertently touch them; the adhesive and dead flies make an unpleasant addition to hair or clothing.

For more information

For a more detailed discussion of how and when to use traps, the North Central Region of the Cooperative Extension Service offers *Alternatives in Insect Management* (NCR 401), a 73-page manual. To order copies, call (217)333-2007.

Prepared with Rick Weinzierl, entomologist, UI Cooperative Extension Service.

DON'T GET ZAPPED BY UNRELIABLE PEST-CONTROL DEVICES

Many products that claim to combat insects with a minimal effort on your part are likely to produce minimal results. Consider the following examples...

Bug zappers

The relatively common "bug zappers" that you see in backyards—lights with electrocuting grids—are at best ineffective. Sometimes, they can actually increase the number of undesirable insects in a backyard. Their light attracts insects from the surrounding area, and many of these insects enter your yard but don't fly into the zapper. Objective studies consistently find more pest insects in yards with bug zappers than in yards without them!

Another drawback is that bug zappers kill a lot of beneficial insects—green lacewings, predators of aphids, and other insects that help control pests.

It should be noted, however, that bug zappers are somewhat effective when used in grocery stores, restaurants, and similar businesses. If the zappers are not visible to insects outdoors, they can help to control insects that have already entered the building.

Ultrasonic pest repellers

According to claims, these devices produce high-frequency sound waves that drive away rodents and insects.

A slight amount of evidence has shown that rodents are temporarily repelled by the ultrasonic waves emitted from some of these devices, but the rodents often adjust. Also, the ultrasonic repellers cover only a very small area of a home and do not send sound waves through walls or around corners. In actual use, they are impractical and just plain ineffective.

Excluding pests from your home and keeping your home sanitary (see Chapter 49) may require a bit more work, but such steps are much more effective.

Prepared with Rick Weinzierl, entomologist, UI Cooperative Extension Service.

14 Use pest barriers in your garden

Plant covers

Plant covers were marketed originally as devices that heat the soil and make earlier planting possible. But these covers have a nice bonus: they keep insects away from plants.

To create your own plant covers, attach muslin to a wooden or wire frame and set it over the plants (see illustration). In addition to muslin, other good materials are vented polyethylene, spunbonded polyester, point-bonded polypropylene, and woven plastics.

Some material is so lightweight that you do not need to use a frame or hoops to support it. Simply place these "floating row covers" on the ground, and the plants will lift them up as they grow. Floating row covers work well for sturdy crops that do not grow too tall, such as cucumbers, gourds, muskmelons, pumpkins, squash, and watermelon.

Most gardeners use plant covers only when plants are small. As plants grow larger, covers are usually removed to allow for weeding and, more importantly, pollination.

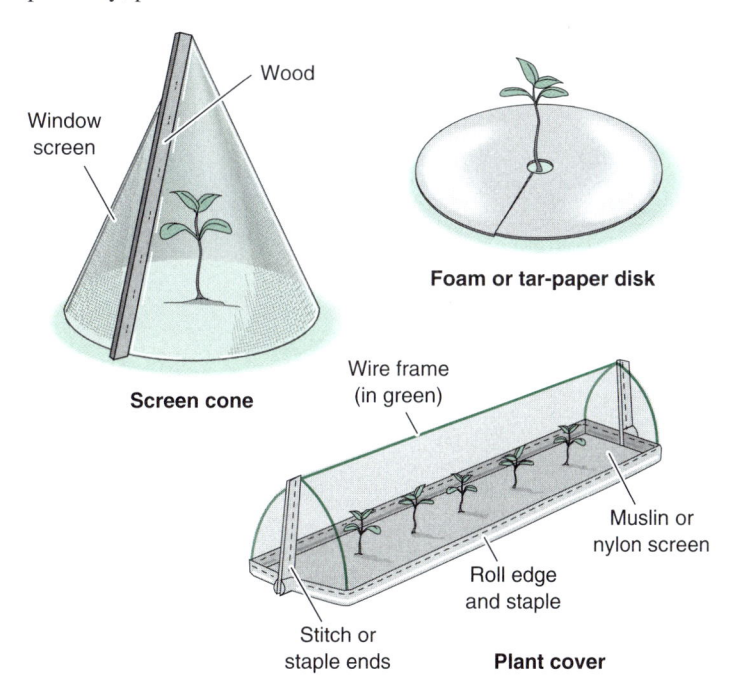

Plant covers provide effective insect barriers when crops are small. In addition, screen cones and foam or tar-paper disks are good for thwarting cabbage maggots.

Other pest barriers

A variety of other barriers are available for use in the yard and garden, many of them specifically targeted to certain pests. Here are a few examples:

Copper sheeting. Place 6-inch-wide copper sheeting at least 2 inches into the soil around valuable plantings. This barrier is quite expensive, but it effectively blocks the invasion of slugs.

Sticky traps. Various sticky barriers and trapping adhesives are available commercially. By attaching or applying them around the trunks of trees, you can create a barrier for canker worms and protect individual trees from defoliation by these particular pests.

Screen cones. To keep cabbage maggots from attacking the roots of cabbage, broccoli, cauliflower, brussels sprouts, radishes, turnips, and related vegetables, create a screen cone from ordinary window screening. Shape the screening into a cone, attach it to a piece of wood, and place it over the plant (see illustration, page 77). This barrier excludes the flies that lay the eggs from which cabbage maggots hatch.

Foam or tar-paper disks. You can also protect plants from cabbage maggots by encircling the young plant with a 3-inch-diameter disk of tar paper, foam rubber, or other sturdy material (see illustration). It works in a manner similar to that of screen cones.

Prepared with Rick Weinzierl, entomologist, UI Cooperative Extension Service.

15 Solarize soil

Putting on the heat

An effective way to suppress several soil pest problems is to moisten and heat the soil. Done properly, this heating will kill or suppress many fungi, nematodes, bacteria, weeds, and insects.

One procedure, known as "solarizing" the soil, consists of placing a transparent plastic tarp on the soil surface. The plastic tarp uses the heat of the sun to raise the soil temperature.

Solarizing is most effective when done for four to six weeks during the hottest parts of the year. Because you cannot solarize the soil when a crop is growing, consider solarizing separate portions of your garden at different times—either in late spring before you plant a late crop or in midsummer to late summer in areas of the garden where you harvested an early crop.

The procedure

To solarize soil in vegetable gardens or planting beds, follow these steps:

1. Level the soil. Remove weeds, plants, and crop debris, and break up large clods of soil.
2. Wet the soil, because moisture helps the heat to penetrate the ground and makes pests more sensitive to high temperatures. If you are covering a small area in a garden, wet the soil *before* putting down the tarp. If you are covering a large area, it may be more practical to wet the soil *after* putting down the tarp. To do so, insert a hose under one end of the tarp, or use trickle irrigation, in which water seeps out from the sides of specially designed hoses.
3. Place the plastic tarp on the soil, eliminating air pockets. To anchor the tarp, bury the edges with soil.
4. Leave the plastic in place for four to six weeks in full summer sun. If you leave the plastic in place any longer, the tarp may become brittle, may tear, and may be difficult to remove.

What kind of plastic tarp is best?

The tarp should be transparent because black or colored tarps will not heat the soil sufficiently. The ideal thickness for the tarp is 1 mil, although a tarp this thin can sometimes tear easily, especially in windy areas. If wind is a problem, use tarps that are 1.5 to 2 mils thick, but avoid 4- to 6-mil tarps.

Using strips of plastic

When solarizing small areas, you may find it easier to cover only the planting beds with strips of plastic, leaving between-row or between-bed furrows uncovered. If you solarize with strips of plastic, it is most effective if the planting beds are 48 to 60 inches wide and oriented north to south.

Prepared with Darin Eastburn, plant pathologist, and Rick Weinzierl, entomologist, UI Cooperative Extension Service.

16 Consider the potential of natural fertilizers

Pros and cons

From the perspective of a plant, it doesn't make any difference whether the form of nitrogen that it uses comes from a traditional concentrated fertilizer or from an organic fertilizer such as manure, dried blood, or fish meal. But there are some advantages to organic fertilizer:

- Organic fertilizer improves soil "tilth"—the condition, or health, of the soil.
- It is less likely to burn the lawn.
- It is less likely to leach, or move down, through the soil and into groundwater.
- It provides a larger complement of minerals for plant growth.

Along with the advantages, however, there are disadvantages.

Because natural fertilizers provide relatively low amounts of nutrients, as compared to concentrated fertilizers such as 10-10-10, it takes a greater volume of organic material to provide the nutrients your plants require. This means it will likely take more effort to incorporate natural fertilizers in your garden.

It also takes time for bacteria and fungi to break down organic fertilizer into the nutrients that plants require. The nutrients provided by many traditional concentrated fertilizers, on the other hand, usually are available as soon as they leach into the soil or are added to it. As a result, organic gardening requires that you plan ahead to meet the nutritional needs of your plants.

Many of the organic fertilizers described in this chapter are available at garden supply stores. However, organic fertilizers tend to be more expensive than traditional products.

Know your soil

Your success with organic fertilizer depends on knowing two things: the fertility of your soil and the nutrient value of the organic matter you intend to use as fertilizer.

Vegetables prefer to grow in fertile, well-drained, "loose" soil with a pH between 6.0 and 6.9. Essential mineral elements will not be as readily available for plant growth if the soil pH is below 6 or above 7.

A simple soil test, performed by a reputable laboratory, can determine the pH level of your soil. This will tell you whether you need to raise or lower the pH, add key nutrients, or do a combination of the two.

The most important nutrients

Of the nutrients that plants need, nitrogen, phosphorus, and potassium are the most important ones that you can add to the soil to increase fertility.

Once you have your soil tested and bring nutrient levels up to adequate levels, you will need to add supplemental nutrients each year to replace the nitrogen, phosphorus, and potassium extracted by fast-growing vegetables. The following are some of the organic fertilizers that provide these key nutrients. Be sure to follow the application rates recommended on the product label.

Nitrogen

Organic fertilizers such as manure, dried blood, fish emulsion, or cottonseed meal can replace the nitrogen used by plants if you apply adequate quantities each year.

Because organic fertilizers do not release nutrients immediately, apply them at planting time so nitrogen will be available to the vegetables during their peak nitrogen demand—the period of maximum plant growth, from three to ten weeks after seeding.

Here are the most common organic sources of nitrogen:

Horse or cattle manure. One of the oldest fertilizers, fresh manure (other than that from poultry) normally will contain 0.5 percent nitrogen. This is $1/20$ of the nitrogen found in an equal amount of 10-10-10 fertilizer, so it takes about twenty times more manure to supply the same amount of nitrogen.

Most of the available nitrogen is in the urine, so it is best to apply fresh manure to the soil about two weeks before planting; that way, you will not injure tender seedlings by exposing them to concentrated urine.

Poultry manure. This often contains three to four times more nitrogen than cow manure, so use less to avoid burning plants with too much nitrogen. Allow manure to dry before applying it to your garden.

Dried blood. This is available in a fine powder. Although blood from slaughterhouses is relatively expensive, it is a good source of quickly available nitrogen.

Fish meal or fish wastes. Although these contain some nitrogen, the nitrogen is available to plants only slowly; thus, plant growth may suffer. Fish emulsion is a different product that contains up to 5 percent readily available nitrogen. Supplying plants with adequate nitrogen solely from fish emulsion can be quite expensive.

Cottonseed meal. This is an expensive byproduct of the cotton industry that contains some nitrogen. The nitrogen it supplies is not readily available, so don't rely on the meal as a sole source of nitrogen for vegetables.

Phosphorus

Most Midwest soils contain some phosphorus. However, the amount may be inadequate to support rapidly growing vegetables. Also, phosphorus is quickly "tied up" when the soil pH is below 5.5 (acidic soil) or above 7.5 (alkaline soil). The product label will explain how much organic fertilizer to apply each year to maintain an adequate level of phosphorus. Sulfur should be added with steamed bone meal and ground rock phosphate to maintain the soil's proper pH.

Here are the most common organic sources of phosphorus:

Steamed bone meal. A byproduct of slaughterhouses, this is a relatively good source of phosphorus, but it can be expensive.

Rock phosphate. This ground rock releases phosphorus very slowly. Although laboratory tests indicate that rock phosphate contains as much as 30 percent phosphorus, it is not in an available form, so it will do little good, especially in alkaline soils.

Horse, cattle, or poultry manure. In addition to supplying nitrogen, manure supplies a fair amount of phosphorus.

Potassium

Before carrying large quantities of organic potassium to your garden, have the soil tested. If the potassium level is above 300 pounds per acre, no additional potassium is required. If it is below that, use one of the following organic potassium sources:

Granite dust. This contains about 5 percent potassium, little of which will be available for crop growth. Finely ground rock simply will not supply elements fast enough for vegetables.

Greensand. This is mined from sedimentary deposits of rock that contain about 6 percent potassium. The potassium is available very slowly to plants.

Wood ash. This is an excellent source of potassium that is rapidly available yet is not likely to burn plants. Do not use charcoal ash, however. It may contain toxic amounts of sulfur.

Horse, cattle, and poultry manure. In addition to supplying nitrogen, manure supplies a fair amount of potassium.

Soil building

No fertilizer will alleviate poor soil-building practices. So, be sure to supply your garden with generous amounts of compost to maintain organic matter in the soil and promote good plant growth.

Prepared with James Schmidt, horticulturist, and Thomas Voigt, turfgrass specialist, UI Cooperative Extension Service.

Using chemicals on the landscape—if necessary

The question may have occurred to you: What's a section on the use of chemicals doing in a book about protecting the home environment?

Although some people prefer to never use synthetic pesticides or fertilizers, others argue that the risks they pose are generally acceptable. Where the use of these chemicals is necessary (and this decision will vary among individuals), pesticides and fertilizers must be handled appropriately to reduce their risks to your health and the environment.

Here are some ideas on how to do just that.

17 Know what to ask a lawn-care company and arborist

Working with a lawn-care company

If you select and work with a lawn-care company that will use pesticides and fertilizers to maintain your lawn, keep these points in mind:

- Make sure the lawn-care company has a state license for the application of pesticides.
- If you prefer minimal use of synthetic pesticides, ask the company about the alternatives they provide.
- When you hire a lawn-care company, you are hiring a *service*. Companies should be able to respond within forty-eight hours of most requests.
- When you hire a lawn-care company, you enter a partnership. Don't expect the company to do 100 percent of the work. For a lawn management program to work, you must mow and water according to recommendations.
- Make sure the company surveys the lawn and prepares a management plan for you ahead of time. Also, be aware that some companies may suggest more applications than are really necessary. To obtain a typical lawn-care schedule for your area, contact your nearest Cooperative Extension Service office.
- Make sure the company tells you ahead of time when applications will be made. They should give you prior warning so you can clear the yard of toys, grills, pets, and other items.
- Ask the applicators which problems they can control and which they cannot. The company should also give you information about managing the lawn. For instance, some pesticide applications will not work unless you follow them with a thorough watering.
- Ask the company which chemicals they plan to apply. If you wish, they should also be willing to show you a label for each of the chemicals.
- Some companies may have many customers and only a small number of applicators, so they cannot promise that every customer will receive service on the optimum date for certain pesticide and fertilizer applications. Find out how close your application will be to the optimum date.
- Stay off a treated lawn until chemicals have dried on the foliage. If you feel more comfortable waiting even longer, then do so.

Most lawn-care companies will post signs indicating that chemicals have been applied; the sign should also indicate how long you need to stay off the lawn.

What about IPM and organic programs?

Lawn-care companies have traditionally relied on preventive applications—putting down chemicals before a pest problem arises. These applications, often made according to a calendar schedule, act as a form of pest-control "insurance." The drawback is that some of the chemical applications may be unnecessary.

One alternative is to look for a company that checks for pest problems and applies chemicals only when the potential exists for unacceptable levels of damage. However, finding such a company can be difficult unless you live in a large metropolitan area; most lawn-care companies offer contracts that call for regular, calendar-date applications that ensure them a steady income.

Some lawn-care companies are beginning to provide programs in which they apply organic fertilizers. Be aware, however, that organic fertilizers do not work as fast as synthetic fertilizers; for the most part, they are slow-release fertilizers that do not give an immediate flush of green growth. Therefore, if you give an organic program only a one-year tryout, you may be disappointed. It usually takes two or three seasons for organic fertilizers to produce optimum results. But in the long run, they may improve the soil's tilth (physical condition).

Most lawn-care companies do not offer organic weed and insect control programs, although some may combine chemical and nonchemical approaches. Currently, an organic program would not produce the same "quality" of turf that results from management programs using synthetic chemicals.

Also, be aware that some companies may use the "organic" name to attract customers but may offer a program that is not much different than traditional pest control. Ask for details.

Working with arborists

You may need to consult with an arborist before cutting tree roots to put in a new sidewalk or before doing other construction work. People often hire arborists to do pruning also, which may be necessary to shape trees, control their size, thin branches, provide better structure, reduce wind resistance, or prevent branches from interfering with wires. Arborists can also help by fertilizing, providing lightning protection systems, and installing supports between forked trunks and branches.

Some homeowners hire arborists to check on their trees once or twice a year. This can be a good approach, for it catches problems before they do irreparable damage. But it is expensive.

When hiring an arborist, keep these tips in mind:

- Beware of "door knockers," especially after storms when nonprofessionals are looking for an easy way to make quick money. Most reputable arborists have all the work they can handle without going door to door.
- Don't be rushed, even if a company says they will give you a 10 percent discount for a decision today. Do *not* pay in advance.
- Ask for certificates of insurance, including proof of liability for personal and property damage and worker's compensation. Then phone the insurance company to make sure the policy is current. In some cases, you may be held financially responsible if an uninsured worker is hurt on your property or damage is done to a neighbor's property.
- Ask for local references—other jobs that the company or individual has done. Then look at these jobs and, if possible, talk to the former client.
- Determine if the arborist is a member of the American Forestry Association, the American Society of Consulting Arborists, the International Society of Arboriculture, or the National Arborist Association. Membership does not guarantee quality, but a lack of membership can cast doubts on the person's professionalism.
- If arborist certification is available in your area, determine whether the arborist complies with the requirements. The International Society of Arboriculture can give you a list of certified arborists where you live. You can reach them at P.O. Box GG, 6 Dunlap Ct., Savoy, IL 61874, telephone: (217)355-9411. Keep in mind, however, that just as with membership in professional associations, certification does not guarantee quality.
- A good arborist will offer a wide range of services: pruning, fertilizing, cabling or bracing, protecting against lightning, and repairing damage.
- A good arborist will recommend "topping" a tree only under very rare circumstances—to save the tree after severe damage to its crown. "Topping" means giving the tree a flat-top appearance by pruning back all of the upper limbs.
- If the arborist recommends tree injections of pesticide or other pesticide applications, he or she must have a valid pesticide applicator's license.

- A conscientious arborist will not use climbing spikes unless the tree is being removed.
- Tree removal is a last resort. Beware of an arborist who is eager to remove a living tree.
- To save money, get together with neighbors and ask about a group discount.
- Find out exactly what work is to be done and how much you will be charged. Work may be done for a single price or for an hourly fee plus the cost of materials. If an arborist uses the latter pricing method, make sure you understand what the maximum charge would be. Obtain an estimate in writing.
- Make sure the contract specifies the dates that work will begin and end.
- Find out what cleanup work will be done and when. Decide who gets any firewood or wood chips. If you will get the firewood, will it be cut into 16-inch lengths and stacked by the garage?
- Find out if tree removal includes grinding out the stump and surface roots to 1 foot below grade, filling the hole with topsoil, and planting grass.

Prepared with Thomas Voigt, turfgrass specialist, Patrick Weicherding, urban forestry specialist, and David Williams, woody ornamentals specialist, UI Cooperative Extension Service.

Additional source: How to Hire an Arborist, *Tree City USA Bulletin No.6, National Arbor Day Foundation,100 Arbor Ave., Nebraska City, NE 68410.*

Jack Robertson

Lawn-care industry moves toward less toxic products

It may come as some surprise to know that homeowners apply more pesticides *per acre* than farmers, even though farmers apply a greater volume of pesticides overall. But things are changing on the home front.

"We are applying about half of the product that we used when we started eighteen years ago," says Jack Robertson, owner of Robertson's Lawn Care in Springfield, Illinois. "In addition, the way the industry must go—and is going—is to less toxic products. So you win both ways."

For example, Robertson says his company has shifted to Barricade, a less toxic crabgrass herbicide that has been available for a couple of years. And in 1994, a new, lower toxicity insecticide called Merit arrived on the market to combat grubs—a major lawn pest.

Improvements in grass seed have also led to varieties that are more resistant to disease, drought, and insects, he adds. This resistance means fewer pest problems and less pesticide use.

If you do wind up with pest problems, Robertson says a key to reducing chemical use is accurate timing.

"Timing is everything," he says. "Stores sell pest control products any time of the year, which is great. But the problem is: Who is there to tell customers the right time to apply? Without proper timing, the chemicals can be a pure waste."

The mode of application has a major impact on chemical use as well, Robertson continues. For example, he says his company tries to spot-treat broadleaf weeds, rather than make a general application across an entire lawn.

"Some lawns may warrant a general broadleaf weed application, but your average lawn does not," he says.

If you want a healthy lawn—a lawn that is more resistant to pest problems—Robertson stresses that homeowners should not expect the lawn-care company to do everything. As he puts it, "Lawn care has three components—the homeowner, the company, and Mother Nature. We all need to work together."

In other words, even with professional help, accomplishing the simple cultural practices of mowing, watering, seeding, fertilizing, and aerating usually are your responsibility. These practices help to create a dense lawn. And according to Robertson, "A dense lawn is your best defense against weeds."

Robertson goes on to note that his company will apply organic fertilizers when customers request it. But only a few customers choose organic fertilizer, he says, because the cost is somewhat higher. Robertson says you can get a good-quality lawn with organic fertilizers; just keep in mind that they don't act as quickly as synthetic fertilizers.

He offers this additional advice on choosing lawn-care professionals:

- Deal with a company that has a good reputation with your neighbors.

- Ask if the company has a full-time office staff that can answer your questions as they arise.

- If the prices seem too low, ask why. "It's like anything else," Robertson says. "You're going to get what you pay for." A homeowner with an average-sized lawn of about 5,000 square feet can expect to pay $30 to $40 per chemical application, he says.

- Finally, check to see if the company is registered with a professional organization. Robertson recommends the Professional Lawn Care Association of America in Marietta, Georgia, (800)458-3466.

18 Understand pesticide toxicity

Pesticides as poisons

Numerous household products are toxic, but pesticides are among the few products used intentionally as poisons. Therefore, it is important to be familiar with the toxicity of these chemicals—no matter whether you plan to use natural or synthetic pesticides.

Natural pesticides, which include plant extracts, viruses, or bacteria, are generally safer than synthetic pesticides, which are developed by people. However, natural pesticides may be toxic nonetheless—and some of them are just as toxic as certain synthetic pesticides, maybe even more so.

Toxicity is the inherent ability of a pesticide to cause injury or death, indicating how poisonous the chemical is. It is usually described in two ways:

Acute toxicity. The ability of a substance to cause harm as a result of a *single dose or exposure* to a chemical.

Chronic toxicity. The ability of a substance to cause harm as the result of *repeated doses or exposures* over time.

Acute toxicity

The acute toxicity of a pesticide is generally described in terms of an LD_{50} value—the dose required to kill 50 percent of the test animals in a laboratory experiment (LD is the abbreviation for "lethal dose"). The *lower* the LD_{50} value, the *higher* the toxicity. For example, a pesticide described by an LD_{50} value of 60 milligrams per kilogram (mg/kg) of test animal body weight is much more toxic than one described by an LD_{50} value of 5,000 mg/kg.

To determine LD_{50} values, researchers expose groups of test animals—usually rats and mice—to a range of single doses. From the results, they calculate LD_{50} values, most often for oral exposure (through the mouth) and for dermal exposure (through the skin).

LD_{50} values are important measures of toxicity, but keep in mind that they do not present a complete picture. They indicate the likelihood of death occurring from a single dose, but they do not indicate the potential for the chemical to cause other acute problems, such as eye injury, throat and lung irritation, chemical burns, or neurological damage. The potential for an individual pesticide to cause injury is also influenced by the length of exposure, how much it is diluted, and the other chemicals with which it is combined.

Chronic toxicity

The chronic effects of chemicals are much harder to pin down than the acute effects. To investigate chronic effects, researchers administer a pesticide to lab animals on a regular basis (often daily) at a range of doses, some very high. The animals are later examined for tumors and other changes in major tissues and organs. Investigators also look at reproductive effects, including impacts on fertility, birth weight, and the incidence of birth defects.

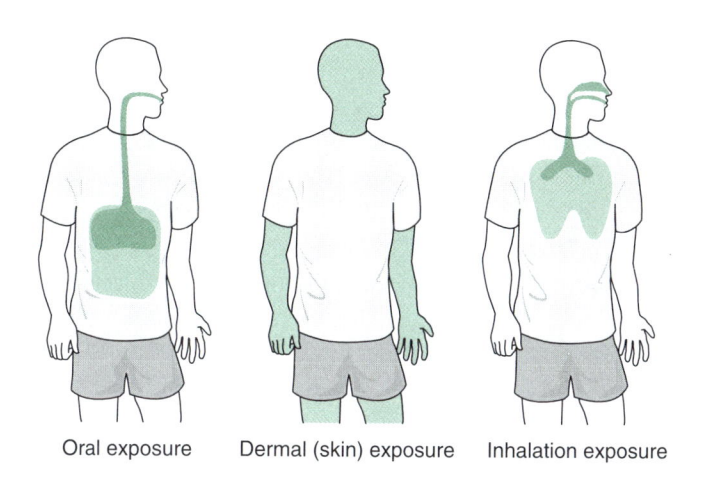

Oral exposure   Dermal (skin) exposure   Inhalation exposure

Pesticides can enter your body by three main routes: through the mouth, through the skin, and through the lungs.

How valid are these tests?

The tests for acute and chronic toxicity are the only scientifically based methods currently used to predict risks to users and consumers. They face these limitations:

- Different species react to chemicals in different ways. Therefore, tests that measure the effects of chemicals on rodents may not always accurately predict effects on humans.
- The tests look at the impact of one compound. They do not take into consideration the possible interaction and effect of several compounds together.
- In studying the long-term, chronic effects of a chemical, researchers must give high doses to test animals. This makes it very difficult to predict the effect of infrequent, *low* levels of exposures—the types of exposures that most people face. Lower exposure usually means lower risk.

Despite these limitations, toxicological studies allow researchers and consumers to make meaningful comparisons among both synthetic and natural pesticides.

> ### ✓ CHECK IT OUT
> For details on natural pesticides: pages 56, 62.

How can you determine the toxicity of a chemical?

Most pesticide labels do not list the chemical's LD_{50} value. However, labels do include "signal words," which give an idea of the pesticide's toxicity or corrosiveness (ability to damage the eyes, nasal passages, or similar tissues). Labels may also warn of possible environmental hazards. Look for one of these signal words:

DANGER-Poison. This indicates a highly toxic pesticide. It may take a few drops to 1 teaspoon of the chemical in concentrated form, taken through the mouth, to kill an adult. Severe injury may occur at even lower doses.

DANGER. This indicates a highly corrosive pesticide. Although the pesticide may not be extremely toxic, it may severely injure the eyes or respiratory and digestive tracts. It may also cause severe skin burns.

Warning. This indicates a moderately toxic or corrosive pesticide. On average, it takes 1 teaspoon to 1 ounce of the chemical in concentrated form, taken through the mouth, to kill an adult. Again, severe injury may occur at lower doses.

Caution. This indicates relatively low toxicity or corrosiveness. It takes more than 1 ounce in a single dose of the concentrated chemical to kill the "average" adult.

Keep in mind that these signal words are relative terms. They indicate how different pesticides compare to one another. Even a pesticide considered to be relatively low in toxicity and labeled with the signal word Caution *can be a deadly poison at a fairly low dose.*

If you want to find out information on the toxicity of a specific pesticide, request a material safety data sheet from the chemical dealer. Manufacturers are required to provide material safety data sheets, which list additional health and safety information.

Prepared with Rick Weinzierl, entomologist, Rhonda Ferree, pesticide applicator training specialist, and Philip Nixon, entomologist, UI Cooperative Extension Service; and Christine Wagner-Hulme, UI graduate research assistant.

A PESTICIDE PRIMER

Types of pesticides

Pesticide is a broad term that describes chemicals used to control many types of plants, animals, and fungi. Among the various types of pesticides, homeowners are most likely to use four:

Herbicides	Used to control weeds
Insecticides	Used to control insects
Rodenticides	Used to control rodents
Fungicides	Used to control fungi

Pesticide names

When trying to find information about a specific pesticide product, keep in mind that it can go by at least two different names:

Generic name — The name of the chemical compound, such as carbaryl. If you're looking for information about a pesticide, it is most effective to search by this name.

Product name — This is the product's brand name. Carbaryl, for instance, can be sold under a variety of brand names, including Sevin, a common insecticide found in garden supply stores.

Pesticide formulations

Finally, be aware that pesticide products are sold in different forms, or "formulations." Here are some of the most common:

Bait — Pesticide mixed with food or other bait.

Dust — Pesticide mixed with talc or a similar dry material. Dusts are applied in the dry form.

Fumigant — Poison gas.

Granule — Pesticide combined with a carrier to form granules. Applied in dry form.

Sprays

Aerosol	Pesticide sprayed from a pressurized can.
Emulsifiable concentrate	A petroleum-based liquid pesticide that has been combined with emulsifiers. Emulsifiers enable you to mix the chemical with water before spraying.
Microencapsulated material	Pesticide with a plastic coating. Mixed with water and sprayed. As the coating breaks down, pesticide is released.
Water-soluble concentrate	Liquid pesticide that you dissolve in water before spraying.
Wettable powder	Pesticide combined with mineral clay and ground into a fine powder. You mix the powder with water, then spray.

Prepared with Rhonda Ferree, pesticide applicator training specialist, and Rick Weinzierl, entomologist, UI Cooperative Extension Service.

19 Read the pesticide label

Read a good label lately?

The most valuable time spent in pest control is the time you take to read the pesticide label.

Before you *buy* a pesticide, read the label to determine:
- Whether it is the right pesticide for the job.
- Whether the pesticide can be used safely under your application conditions.
- Whether there are any restrictions on the pesticide.
- How much pesticide you should buy for the area you are treating.
- When to apply the pesticide.

Before you *mix and apply* a pesticide, read the label to determine:
- What protective clothing to use and safety measures to follow.
- What the chemical can be mixed with.
- How much pesticide to mix.
- The mixing procedure.
- How to apply the pesticide.
- How long you should wait after application to reenter the area, harvest the crop, or plant another crop.

Before you *store or dispose* of a pesticide, read the label to determine:
- Where and how to store the pesticide.
- How to clean and dispose of the chemical container.
- How to dispose of surplus pesticide.

Refer to the accompanying sample label to find out where to locate this information.

Material safety data sheets

In addition to meeting labeling requirements, chemical manufacturers must provide material safety data sheets through dealers and other chemical distributors. If you want to find out additional health and safety information about a particular product, ask for a material safety data sheet from the chemical dealer.

The label

The following sample label is for a microbial insecticide known as *Bacillus thuringiensis*, or *Bt*. (See Chapter 10 for details on microbial insecticides.) Although *Bt* is a less toxic alternative to more traditional, synthetic pesticides, its label includes the key elements found on any label.

Formulation. The label tells you the product's formulation—such as spray, dust, granule, or fumigant.

Active ingredients. The label must list the active ingredient—the ingredient that actually kills or inhibits the pest.

Inert ingredients. Inert ingredients (such as carriers or solvents) do not have to be specified, but their concentration must be listed.

Level of toxicity. The label indicates the level of toxicity with one of three signal words: Danger-poison (high toxicity), Warning (moderate toxicity), or Caution (low or slight toxicity).

Precautionary statements. The label describes the hazards to the applicator, children, domestic animals, wildlife, and the environment. This section is usually much longer for more potent, synthetic pesticides than it is for a biological insecticide such as the one below. If protective clothing and equipment are necessary, the precautionary statements will tell you.

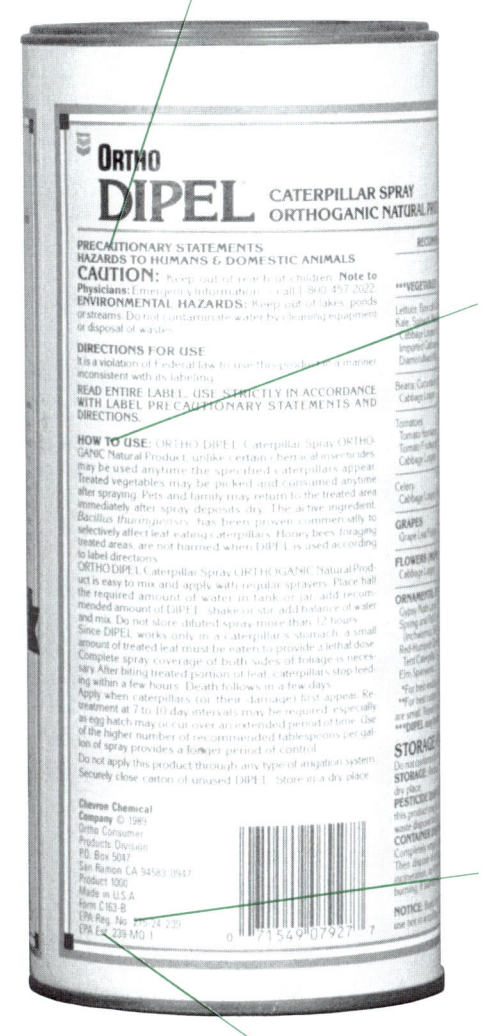

Directions for use. The directions tell you such information as: how to mix the pesticide; when to apply the pesticide; when you can return to the treated area after application; when you can safely harvest and consume the crop after application; and what application rate to use. Application rates are typically grouped by the various crops and areas for which the pesticide may legally be used.

EPA registration number. The registration number shows that the chemical has been registered with the U.S. EPA.

Establishment number. The establishment number identifies the facility that produced the pesticide.

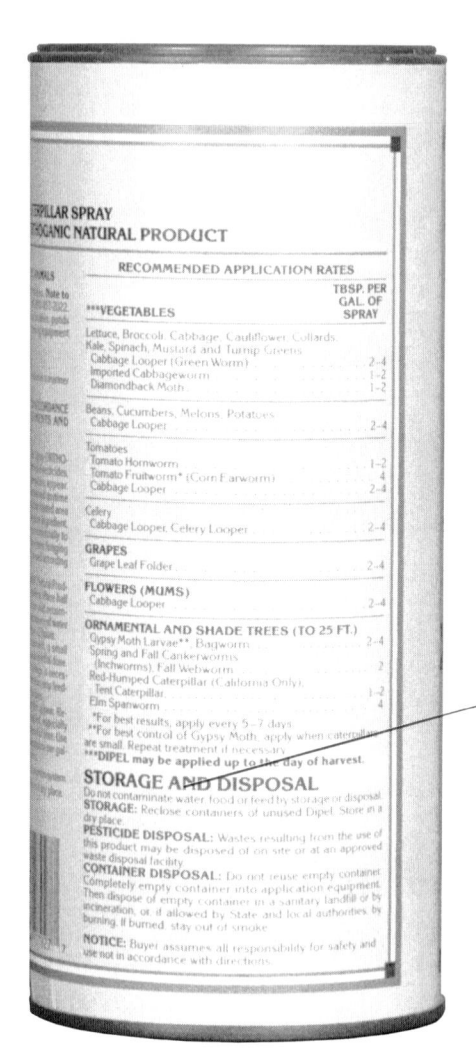

Storage and disposal. The label tells you how to dispose of both the pesticide and the container. Some pesticide labels also tell you what to do if you spill the product.

Prepared with Rhonda Ferree, pesticide applicator training specialist, and Rick Weinzierl, entomologist, UI Cooperative Extension Service.

20 Apply pesticides accurately and safely

Overapplication is overkill

More is not better when it comes to applying pesticides. Applying too much pesticide can damage your lawn, plants, or whatever else you are treating. In addition, overapplication increases your exposure to chemicals, is a waste of money, and can harm the environment. To understand how you can apply pesticides with the least amount of waste and the least risk of exposure, you first need to understand the basic types of application equipment.

Pesticide application equipment

Hose-end sprayers. A hose-end sprayer is a relatively simple and inexpensive sprayer. It consists of a jar and lid mechanism with a nozzle on one end and a hose connection on the other. As water from your hose passes through the nozzle, pesticide solution is drawn out of the jar, mixed with water, and sprayed from the nozzle.

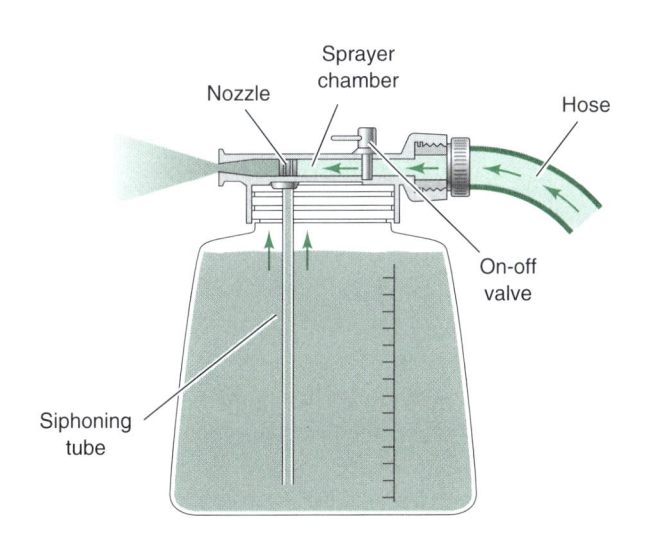

Hose-end sprayer

Manual sprayer

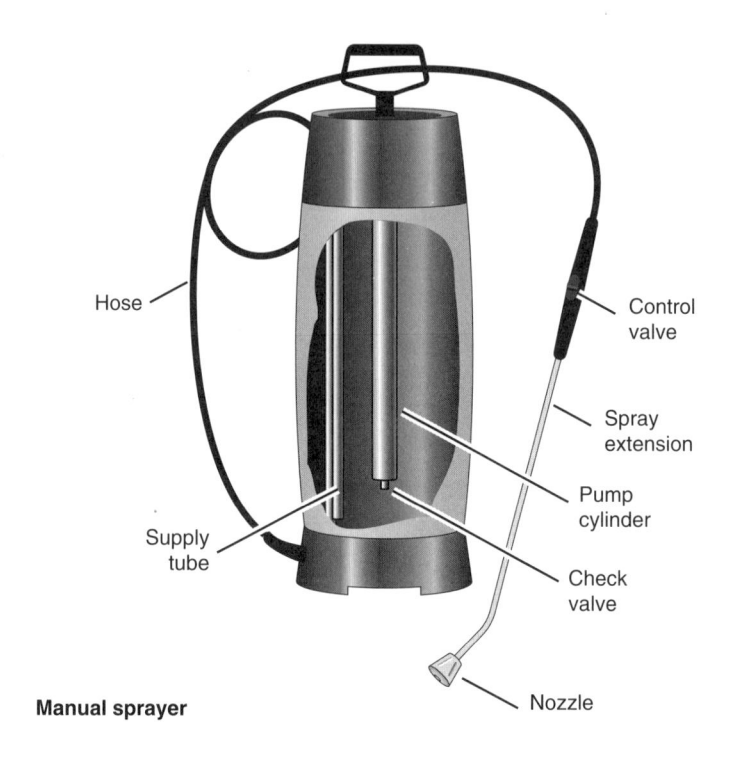

Manual sprayers. Manual sprayers, such as compressed air and knapsack sprayers, are designed for spot treatment and for hard-to-reach areas that cannot be treated with larger units. Unlike hose-end sprayers, manual units do not mix the chemical with water during application. Chemicals must be mixed before use.

Most manual sprayers use compressed air, which forces the spray liquid through a nozzle. Adjustable handguns usually are used with these units. These sprayers are relatively inexpensive, simple to operate, maneuverable, and easy to clean and store.

Trombone sprayers. A hose runs from a trombone sprayer into a bucket containing a chemical mixture. By moving the sprayer's "slide" back and forth—similar to the slide on a trombone—you draw chemical mixture out of the bucket. This kind of sprayer allows you to mix a greater amount of chemical than with the smaller equipment.

Estate sprayers. These sprayers are designed for large lawns. They usually are equipped with a motor-driven pump, have tank capacities of 30 to 50 gallons, and may apply pesticides through a boom. They may also be equipped with a handgun and several feet of hose, which allow you to spray shrubs and trees more conveniently.

Dusters. Equipment to apply pesticide dusts includes squeeze bulbs, bellows, tubes, shakers, or fans powered by a hand crank.

Granular applicators. Granular pesticides often come mixed with fertilizer, but you can also purchase unmixed granular pesticides. Mixed or not, granular pesticides are typically applied with rotary or drop fertilizer spreaders. Drop spreaders drop the granules straight down and are generally more precise than rotary spreaders. Rotary spreaders throw granules in a semicircular arc; therefore, they have a wider swath and can cover a given area faster than drop spreaders. Rotary spreaders include large-capacity units that are pushed on wheels and smaller hand-held devices.

Indoor sprayers. Most indoor pesticides come prepackaged in aerosol containers or hand-pump sprayers. Hand-pump sprayers are nonpressurized containers that you pump to build pressure and apply the product. They often are sold with premixed pesticides.

Calibration

Pesticide labels tell you how much chemical to apply, but these recommendations *do not* and *cannot* take into consideration the wide variation among application equipment. The amount of chemical that one sprayer or spreader puts out differs greatly from the application rate of another brand.

Therefore, to make sure your equipment is applying the correct amount, you need to determine the exact rate at which it applies pesticide. This is called "calibrating" your equipment—a crucial but often ignored procedure for homeowners. The owner's manuals for some sprayers and spreaders provide helpful guidelines on how to set the equipment to obtain the correct application rate—but they are just estimates. The only way to know for sure is to calibrate the equipment yourself.

For ideas on how to do this, see the sidebar on page 112, "Calibrating your application equipment."

Protective clothing and equipment

Before mixing and applying any pesticide, read the label closely to find out what kind of protective clothing and equipment the product recommends. If a product doesn't mention any protective equipment, that doesn't mean you should completely rule out the need for it.

When using any pesticide, regardless of its toxicity, wear clothing that protects most of your body. What you wear depends on whether you are applying pesticides, mixing them, or cleaning up application equipment.

Protective clothing when applying pesticides. Here are the essentials:

- Hat
- Long-sleeved shirt (with sleeves buttoned down)
- Long-leg trousers or a coverall garment
- Unlined rubber or neoprene gloves. Cloth or leather gloves absorb pesticide, as do cloth linings.
- Socks and unlined boots. Boots are especially important when applying pesticides in wet grass because wet grass increases the chances that chemicals will contaminate your shoes.

Check the label for any additional recommendations on protective clothing, such as goggles, a rubber apron, or a respirator. If the label recommends a respirator, make sure it fits and is maintained properly.

Protective clothing when mixing pesticides. When mixing pesticides, wear the same equipment suggested above for pesticide application. But take extra precautions as well. The pesticide concentrates you handle during mixing are more toxic than the diluted spray, so wearing a respirator, goggles, and rubber apron may be advised.

Protective clothing when cleaning application equipment. When cleaning up pesticide application equipment, wear the same protective gear recommended for pesticide mixing, with one exception—the respirator is usually not necessary.

In addition to wearing protective clothing, observe essential safety measures during the three stages of pesticide use—before and during mixing, before and during spraying, and after spraying.

Before and during mixing

Here are tips for successful mixing:
- Know how large an area you are treating and how much spray solution must be mixed.
- Know where to find emergency response numbers in case of an accident.
- Prepare pesticide mixtures outdoors, where there is plenty of fresh air and good lighting.
- Keep pets, children, food, and dishes away from the area where you are doing the mixing.
- Use measuring cups and tools that have been designated *exclusively* for pesticide use.
- Make sure you have soap, towels, and other supplies available in case of a spill. Cat litter is a good absorbent in the event of a spill.
- When mixing pesticides, keep the sprayer in a containment device, such as a rubber tub. That way, the tub will contain any chemical that spills, preventing environmental contamination. By catching spilled pesticides, you will also be able to salvage the spray for further use.

Before and during spraying

The next step is spraying:
- Read the label thoroughly.
- Make sure your application equipment is in working condition and that the hoses and connections do not leak.
- Inform children and neighbors about your pesticide application plans, and keep pets away from the area. The label should tell you how long you must wait before reentering the treated area. If it doesn't, wait at least until the spray has dried.
- Never eat, drink, or smoke while handling pesticides.
- Prevent drift—the movement of pesticides through the air to any area other than where it was intended. Apply pesticides on a calm day, preferably in the morning when there is less wind. Make your applications as close to the pest as possible and when temperatures

are between 60° and 85°F. At higher temperatures, some pesticides "volatilize"—they change into forms that can more easily drift. Also, do not pump the sprayer too much because the higher pressure increases the chance of drift.
- If you become contaminated while using pesticides, change clothes and shower immediately. Waiting until the end of the day to clean up gives the chemical more time to absorb into the skin.

After spraying

Finishing up:
- Immediately after use, clean the mixing and application equipment inside and out. Wear protective clothing during this task.
- Do not clean equipment directly on the driveway because contaminated rinse water can flow into the sewer. Since most home spraying equipment is small, place it in a rubber tub that has been dedicated for mixing and rinsing procedures. The tub catches rinse water as you thoroughly rinse equipment with a water-detergent solution. Allow the water-detergent to circulate through the system, and run some through the hoses and nozzles. Flush the system twice with clean water, but try to minimize the amount of water used. When you are done, put the captured rinse water in the sprayer and apply it on areas listed on the label. This may leave a little pesticide residue in the sprayer, but that's better than dumping rinse water on the ground.
- After handling pesticides, always wash your hands thoroughly with soap and water. Otherwise, pesticide residues can pass from your hands to anything you touch.
- Change clothes and take a shower every day that you use pesticides.
- Keep contaminated clothing separate from other laundry. Warn the person washing the clothes of possible dangers. If you do not wash pesticide residues from clothing, the residues from even minor chemical exposure can eventually poison the wearer.
- Wash contaminated clothes separately from other laundry in warm or hot water using standard laundry detergents. Clean the washer afterward by running a complete cycle without clothes, but with a full amount of hot water and detergent.
- Dry laundered clothes immediately after washing. Line drying is better than using a clothes dryer.
- If clothes have been contaminated with pesticide *concentrate,* throw them away. They cannot be cleaned properly.
- Read the label to learn of any special storage needs required by the product.
- Dispose of leftover pesticides properly (according to the label).

> **✓ CHECK IT OUT**
> For more information on pesticide disposal: page 121.

Prepared with Rhonda Ferree, pesticide applicator training specialist, and Robert Wolf, agricultural engineer, UI Cooperative Extension Service.

21 Apply fertilizers accurately and safely

Aim for precision

Even lawns can overeat. Fertilizer is food for turfgrass, but too much food can mean trouble. An excessive amount of fertilizer can burn your lawn, and it can leach down through the soil, threatening groundwater. It's also a waste of money.

However, striving for more precise fertilizer applications does not have to be complicated. It requires attention to a few details, beginning with equipment selection.

Drop and rotary spreaders

There are two basic types of granular fertilizer spreaders—drop and rotary—each of which has distinct advantages and disadvantages.

Drop spreaders. A drop spreader does just what its name implies. It drops fertilizer granules straight down. Because the fertilizer falls between the two wheels, you have to overlap each of your passes to cover all the ground.

Drop spreaders, in general, apply fertilizers more precisely and uniformly than rotary spreaders. Because the fertilizer drops straight down, there is less risk of drift and less chance that you will apply granules off

Drop spreader

target. However, some drop spreaders cannot handle large granules, and ground clearance in wet grass can be a problem. Any steering errors with a drop spreader can mean missed strips or areas with excessive overlap—problems that can result in light and dark streaks in the lawn.

Rotary spreaders. Instead of dropping granules straight down, rotary spreaders throw them in a semicircular arc in front of the spreader. Rotary spreaders cover a wider swath and therefore cover a given area faster than drop spreaders. They also are more durable and easier to push than drop spreaders, and steering errors are less critical with them.

On the down side, however, rotary spreaders are less precise in applying and distributing fertilizer. They throw granules in a 6- to 8-foot-wide swath, so the fertilizer often ends up in nontarget areas such as sidewalks, driveways, and patios. The problem becomes more serious when applying fine particles on a windy day; drift can carry the particles even farther.

Rotary spreader

Calibration

Fertilizer labels tell you how much chemical you should apply, but these recommendations do not and *cannot* take into consideration the wide variation among application equipment and granular products. The application rate with one spreader differs greatly from the rate with another spreader.

Therefore, to make sure your equipment is applying the correct amount of fertilizer, you need to determine exactly how much it applies. This is called "calibrating" your equipment. The owner's manual for some spreaders provides helpful guidelines on how to set the equipment to obtain the correct application rate—but these guidelines are just estimates. The only way to be certain how much chemical your spreader applies is to calibrate the equipment yourself.

For ideas on how to do this, see the sidebar on page 112, "Calibrating your application equipment."

Application tips

Whether you have a rotary or a drop spreader, you can avoid excessive application of fertilizer by observing a few simple guidelines:

- Fill the spreader on a paved surface rather than on your lawn. If a spill occurs, it's easy to clean up from a drive.
- When filling the spreader, make sure the openings are closed to prevent leaking on the ground.
- Begin by applying fertilizer along a header strip—an area where you can turn around and realign the spreader at each end of the lawn (see illustration). After making each pass, close the spreader as you move into the header strip. Then turn the spreader around and open the spreader again while moving forward to make the next pass.
- Read the fertilizer label carefully and follow the recommended application rate—the overall amount of product applied in pounds per 1,000 square feet. Your spreader should include instructions on how it can be set for different application rates. For the greatest accuracy, run through the calibration process with each granular product you use.
- Normally, you cannot operate a spreader backward. If you do, rotary spreaders will deliver an uneven pattern, and drop spreaders will release granules at a different rate.
- To spread granules uniformly, maintain a uniform walking speed. If you speed up, the turf will receive less fertilizer, and if you slow down, the turf will receive more fertilizer. Speed will also affect the pattern width of rotary spreaders.

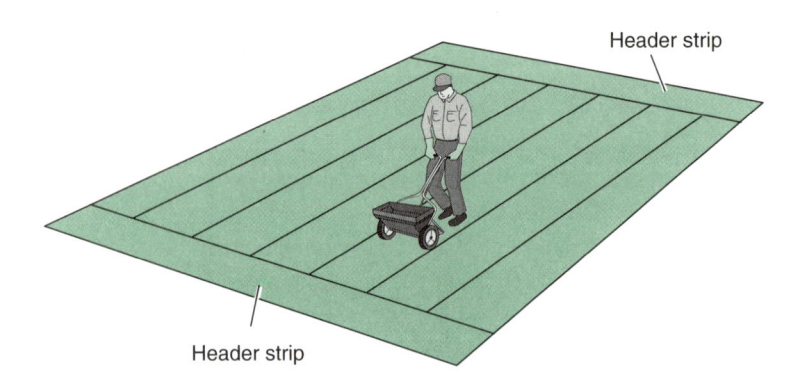

A header strip at each end of the lawn provides an area where you can turn around and realign a spreader.

Some additional tips for rotary spreaders

Because rotary spreaders increase the risk of uneven application, the turf can sometimes appear to have stripes of lighter or darker green, depending on how much fertilizer different areas receive.

To solve this problem, some people will cut the application rate in half and go over the area twice, making the second application at right angles to the first one. But this is *not* a good solution because it will merely change the stripes into a checkerboard. A better solution is to cut the application rate in half and run two applications in the *same* direction, not at right angles to each other. Overlap the rows of each application as shown in the accompanying illustration.

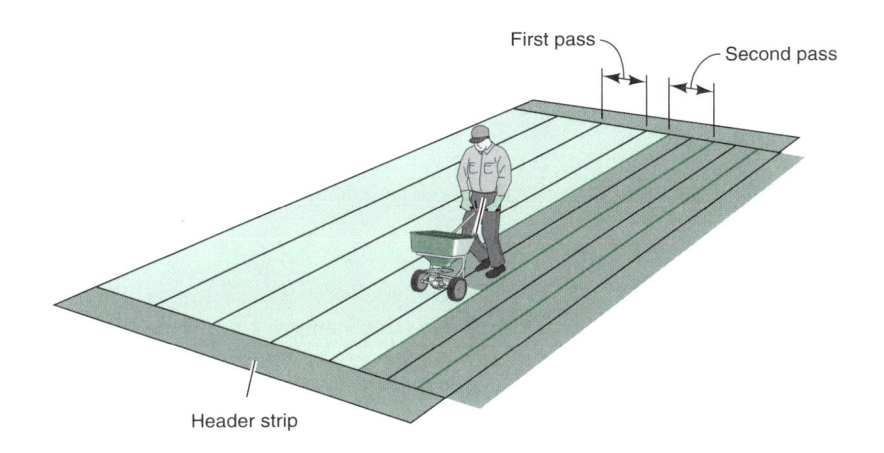

To avoid creating light- and dark-green stripes with a rotary spreader, cut the application rate in half and run two overlapping applications in the same direction. The header strips can be covered before running the two overlapping applications.

Another problem with rotary spreaders is that they throw material 3 to 4 feet in front of the spreader, as well as to the sides. This means you must be extremely careful to avoid tossing fertilizer off target. The problem could be even more serious if the machine is applying a pesticide-fertilizer mixture, such as a "weed 'n' feed" material; pesticides usually do more off-target damage than fertilizer.

To prevent granules from drifting, apply them on a calm day, preferably in the morning when there is less wind.

Prepared with Robert Wolf, agricultural engineer, and David Williams, horticulturist, UI Cooperative Extension Service.

CALIBRATING YOUR APPLICATION EQUIPMENT

Calibrating liquid applicators

Step 1. Mark off an area of 1,000 square feet (20 feet by 50 feet, for instance).

Step 2. Fill the spray canister or spray tank with water, and apply as much as necessary to cover the entire 1,000 square feet, using a proper spray technique. Time how long it took you to spray the area. Record only the amount of time that the sprayer was actually operating.

Step 3. Fill the jar or tank with water again, and spray the water into a container for the amount of time you determined in Step 2.

Step 4. Measure the amount of water that you sprayed into the container. This is the amount of chemical that your sprayer will apply per 1,000 square feet, as long as you use a similar spray pattern during actual application. Record this figure for future reference.

Step 5. Check the label to find out how much chemical to use per gallon. Then mix the necessary amount for the size of area that you are spraying. Spray the chemical mix at the same pace and using the same technique as during calibration.

Example: It took you 50 seconds to spray water on the 1,000-square-foot test course. Therefore, you spray water into a container for 50 seconds and collect 2 gallons. This means the application rate for your equipment is 2 gallons of product per 50 seconds, or per 1,000 square feet. Here is how you then determine the amount of chemical mixture to use:

- The area you are spraying is 6,000 square feet. Because your equipment applies 2 gallons per 1,000 square feet, you need to spray 12 gallons of mixture to cover the entire area.

- The label says to mix ½ ounce of chemical with every gallon of water. Because your sprayer holds 3 gallons, you mix 1½ ounces of chemical with a full sprayer of water.

- To spray 12 gallons of mixture, fill your 3-gallon sprayer four times. By mixing 1½ ounces of chemical with every tankful, you end up using a total of 6 ounces for the entire 6,000 square feet.

Calibrating granular applicators

Step 1. If you are using a *drop spreader,* mark off an area of 1,000 square feet (20 by 50 feet, for instance). If you are using a *rotary spreader,* mark off a 5,000-square-foot area (50 by 100 feet, for instance). It is ideal to do this test on a concrete or asphalt surface so you can sweep up the granules afterward. If you do not have a concrete area large enough, you can mark off an area in the lawn.

Step 2. Weigh a certain amount of granular material (pesticide or fertilizer); then add the granules to the spreader hopper. Adjust the spreader gate to a setting you believe is close to the desired rate. Base this selection on recommendations in your owner's manual and the pesticide label, keeping in mind that it is only a suggested setting; you must calibrate your equipment for each product you use.

Step 3. Apply the granular material to the 1,000-square-foot or 5,000-square-foot area you have marked off.

Step 4. Weigh the amount of material remaining in your spreader.

Step 5. Subtract the amount of material remaining after application from the initial amount of granules in the spreader. The answer is the amount of granular material that your spreader applies per 1,000 or 5,000 square feet. Compare this figure to the rate recommended on the fertilizer label. If the amount is too high or too low, adjust the setting on your spreader accordingly, and repeat steps 2 through 5. When you determine the correct setting, record this information for future reference; but remember, a different product may require a different setting.

Example: Because you are using a drop spreader, you mark off a 1,000-square-foot area. Then you put 6 pounds of fertilizer into the spreader and set the gate to what you believe is the correct setting. After application, you weigh the remaining fertilizer and find that you have 1 pound left in the spreader. You make the following calculation:

6 pounds − 1 pound = 5 pounds per 1,000 square feet

6 pounds is the initial amount of granules in the spreader
1 pound is the amount remaining in the spreader after application
5 pounds is the amount that your spreader applies per 1,000 square feet

The product label says to apply 5 pounds per 1,000 square feet, so you are right on target.

Prepared with Robert Wolf, agricultural engineer, UI Cooperative Extension Service.

Storing and disposing of hazardous chemicals

If you're like many people, you probably have hazardous products that have been sitting on your shelves for as long as some unused wedding presents. In storage, these hazardous products are usually in concentrated form, so they deserve your attention. What follows are some key ideas on how to dispose of these unwanted chemicals, as well as some tips on storing them safely.

22 Store hazardous chemicals safely

The home as a warehouse

When most people think of hazardous materials, they picture trucks full of chemicals from factories, dumps oozing slime, and the byproducts of our synthetic society. Yet every home can be a warehouse of hazardous materials. The U.S. Environmental Protection Agency (EPA) estimates that the average household generates 20 pounds of household hazardous wastes per year and stores up to 100 pounds of hazardous wastes.

To get an idea of the sheer number of chemicals that can pose hazards, especially to small children, consider this partial list of hazardous materials that can be found in the average home:

Acids	Herbicides
Ammonia	Ink
Antifreeze	Insecticides
Aspirin	Iodine
Automotive products	Kerosene
Bathroom bowl cleaner	Lighter fluid
Bleach	Lye
Boric acid	Nail polish
Carbon tetrachloride	Nail polish remover
Cigarettes	Oven cleaner
Cleaning fluid	Paint
Cologne	Paint thinner
Copper and brass cleaner	Perfume
Cosmetics	Permanent wave solution
Dandruff shampoo	Petroleum distillates
Dishwasher detergent	Pine oil
Disinfectants	Plant food
Drain cleaner	Rodenticides
Drugs and medications	Rubbing alcohol
Epoxy glue	Shaving lotion
Fungicides	Silver polish
Furniture polish	Strychnine
Gasoline	Turpentine
Gun cleaner	Vitamins
Hair dye	Window-washing solvent

Classifying hazardous materials

The following are four major classifications of hazardous materials.

Corrosive materials. Corrosive materials can dissolve or wear away other materials. When stored in the wrong container, they can eat through the container. Also, most of the materials that are corrosive to containers are potentially dangerous to the skin and eyes of humans and animals.

A few common corrosives include metal cleaners with phosphoric acid, drain cleaners that contain sulfuric acid, rust removers with hydrofluoric acid, and drain cleaners containing sodium hydroxide or lye.

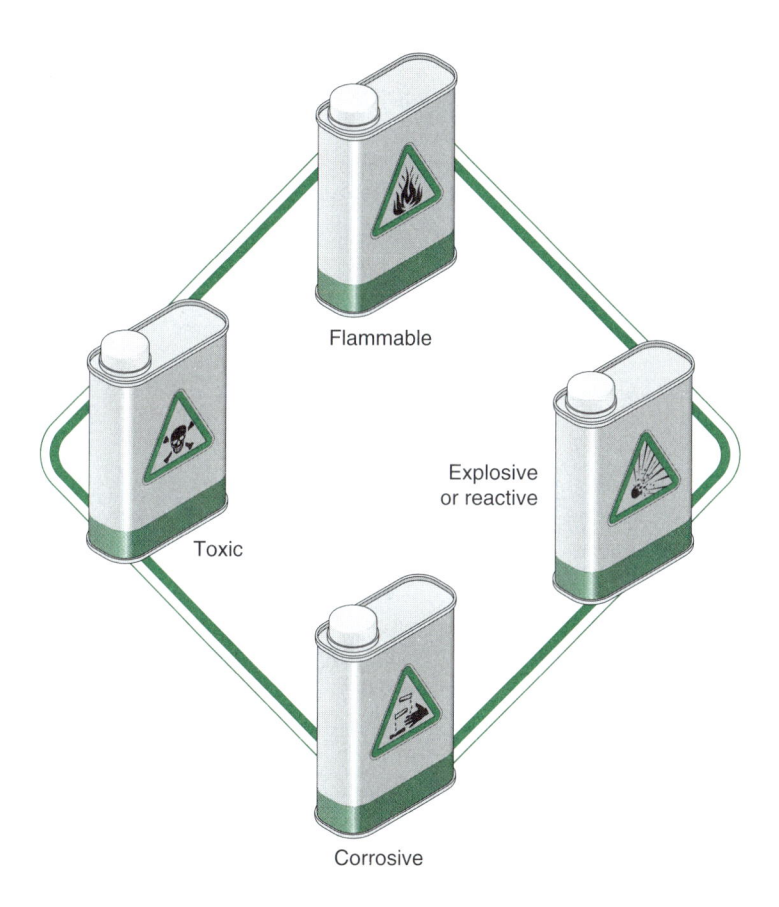

Household hazardous materials fall into one or more of four basic categories: corrosive materials, flammable materials, explosive or reactive materials, and toxic materials.

Flammable materials. Flammable items pose a serious threat of fire if stored improperly. Many of these items indicate, "Do not store near heat" or "Keep in cool, dry place."

Explosive or reactive materials. These materials can explode when combined with other substances. They may also react violently in other ways, such as producing toxic gases. For example, when bleach and many dish detergents that contain chlorine bleach mix with ammonia, lye, or acids, the combination can produce toxic gases.

Toxic materials. Toxic materials are those materials that, in sufficient quantities, pose a hazard to human health. They are sometimes identified with the symbol of a skull and crossbones. Most cupboards and closets are full of potentially toxic materials, from air fresheners and carpet deodorizers to mothballs and oven cleaners.

Toxic materials can often be used safely, however, and they can be beneficial or even necessary to the body, as in the case of some vitamins or medicines. Hazards arise when the amount of material is high enough and the duration of exposure long enough to cause harm.

Many hazardous materials fall into two, three, or all four of these categories. For instance, an acid is corrosive, can be toxic when swallowed, and is reactive when it combines with chlorine.

Household hazardous waste

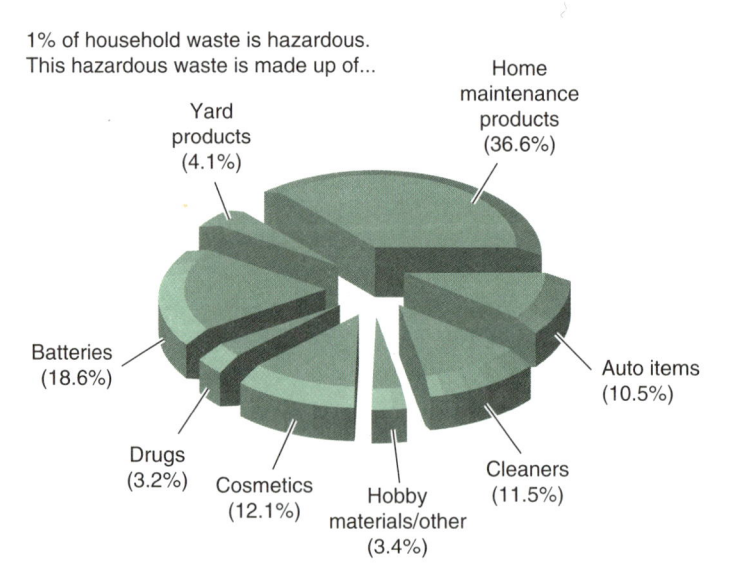

1% of household waste is hazardous. This hazardous waste is made up of...

- Yard products (4.1%)
- Home maintenance products (36.6%)
- Batteries (18.6%)
- Auto items (10.5%)
- Drugs (3.2%)
- Cosmetics (12.1%)
- Hobby materials/other (3.4%)
- Cleaners (11.5%)

Source: Rathje and Wilson, The Garbage Project, University of Arizona.

Storing hazardous materials

- To reduce the amount of hazardous materials in storage, buy only the amount that you need for the job at hand. The chapters that follow explain how to dispose of excess hazardous material.
- Store hazardous materials out of the reach of children.
- Keep hazardous materials in a locked cabinet. This applies to chemicals stored in the garage as well as in the house.
- Buy products with safety closures whenever possible.
- Store hazardous materials in their original containers. This assures that you can identify the product in case of poisoning, and it helps to prevent children from mistaking a poison for something else. It is especially dangerous to store chemicals in containers once used for food or beverages.
- Keep product labels in place. If a label is peeling off, reattach it with transparent tape.
- Do not store hazardous materials near foods or medicines. With the numerous look-alike containers on the market, a hazardous material might be mistaken for food.
- Store purses out of reach of small children. A purse can contain several poisons, such as medicines and cosmetics.
- Do not store pressurized containers (such as aerosols) in the sun, in the car's glove compartment, or near other heat sources that may cause the containers to explode. Also, do not store pressurized containers in wet or damp areas because a can that rusts may rupture or leak.
- Keep a bag of cat litter, sand, or sawdust near the storage area to soak up any spilled chemical from a broken or leaking container. Keep a separate broom and dustpan handy for chemical cleanup, and do not use them for any other purpose.

Added precautions for flammable materials

Fuel. Flammable products such as gasoline, kerosene, propane gas, and paint thinner should be stored in approved containers in the garage—never inside the house. A well-ventilated garage will reduce the risk that vapors given off by flammable liquids might ignite.

Also, when storing fuel, do not fill the container to capacity. By leaving some space in the container, you allow the fuel room to expand during hot weather.

Pesticides. Many liquid pesticides contain a petroleum-based carrier or solvent and therefore pose a fire hazard. Store these pesticides in a garage in a locked cabinet—not in the house. To determine the flammability of a pesticide, check the label.

If a pesticide is flammable and must be stored in a garage, be aware that temperatures below freezing can cause the pesticide to separate from the solvent; in many cases, this makes the chemical ineffective. This problem underscores the importance of buying only the amount of chemical that you need for a job.

Prepared with Rhonda Ferree, pesticide applicator training specialist, Robert Wolf, agricultural engineer, and Robert Aherin, safety specialist, UI Cooperative Extension Service.

23 Dispose of pesticides safely

The problem

Surveys have shown that the most common way homeowners dispose of leftover pesticides is to dump them down the toilet or sink. However, this places a serious burden on the municipality's water-treatment facilities; if a home is connected to a septic system, pesticides can interfere with the system's proper operation.

Also, do not dump pesticides on the ground because doing so can contaminate groundwater. And do not place them in the garbage because they could injure trash collectors and contaminate the environment.

Although the toxicity varies from one product to the next, all pesticides should be treated with caution. Take special care when disposing of these materials:

- Leftover pesticide concentrate
- Leftover pesticide mix (concentrate plus water)
- Rinse water used to clean the sprayer after application
- Empty pesticide containers
- Rinse water used to clean empty pesticide containers

Disposing of leftover pesticide concentrate and mix

Try to purchase only the amount of pesticide that you need for a job so you don't have any leftover chemical. But if you end up with excess pesticide concentrate, dilute it as directed on the label; then apply it to an area listed on the label. You can also dispose of excess *pesticide mix* by applying it to an area listed on the label.

In "using up" a pesticide, however, do not *overuse* it. Do not apply more than is recommended on the label, or you could damage both the environment and the plants you are trying to protect. Overuse also increases your own exposure to these toxic chemicals.

Another option is to store leftover pesticide until you can take it to a hazardous-waste collection site. For ideas on how to store pesticides safely, see Chapter 22.

Disposing of liquid containers and rinse water

An empty pesticide container is not as empty as you might think; a significant amount of pesticide residue can remain inside it. Therefore, you must take certain precautions before you toss an empty container of liquid pesticide into the trash. Follow this triple-rinse procedure:

1. When you are down to the last amount of pesticide concentrate, drain the pesticide container into your spray tank for at least 30 seconds.
2. Fill the empty container one-fifth to one-fourth full of water. Rinse thoroughly.
3. Use this rinse water as dilution water for the pesticide concentrate in the sprayer. If the dilution rate allows you to pour *all* of the rinse water into the sprayer, drain it into the sprayer for at least 30 seconds.
4. Follow the procedure in Steps 2 and 3 two more times.
5. Spray the pesticide mixture on areas listed on the label. Do not exceed the label's application rate.

Triple-rinse containers as soon as they are emptied during the mixing/filling stage. If you wait until later, pesticide residue can dry inside the container, making it difficult to clean with triple-rinsing. Also, be sure to wear protective clothing when triple-rinsing. See Chapter 20 for suggested protection.

If you still have some rinse water left over after using it to dilute pesticide concentrate, spray it on areas listed on the label. You can do the same with rinse water used to clean the sprayer. However, do not exceed the rates specified on the label, and do not dump rinse water down a drain or on the ground. For details on how to properly rinse spray equipment, see page 106.

What about dry pesticide containers?

Although containers that hold liquid pesticides require special attention, bags that hold dry pesticides simply need to be emptied completely before discarding in the trash.

Prepared with Robert Wolf, agricultural engineer, Rhonda Ferree, pesticide applicator training specialist, and Philip Nixon, entomologist, UI Cooperative Extension Service.

24 Dispose of auto products safely

Used oil

The improper disposal of used motor oil can contaminate lakes, rivers, and groundwater. In fact, the used oil from a single oil change can seriously contaminate a *million gallons* of fresh water—enough water to supply fifty people for one year. Used motor oil contains several organic chemicals and metals.

Your best disposal option is to recycle used oil. Contact common collection sites such as local oil distributors, auto-repair stations, and commercial recycling services.

When storing oil before recycling, make sure it is in a clean, sealed container such as a steel drum or plastic jug. Mark the container clearly as used oil and store it away from children. Also, never mix solvents, gasoline, or antifreeze with used motor oil. It is almost impossible to recycle oil once it is contaminated with these products.

Used oil filters

Do-it-yourselfers have become increasingly conscious of recycling used oil. But used oil *filters* deserve attention too. A used oil filter from a passenger car can contain a pint to a quart of oil. And even if a filter is drained overnight, it will still contain 2 to 8 ounces of used oil. That's why some states have banned used oil filters from landfills.

The best disposal option is to "hot-drain" filters and then take them to a household hazardous-waste collection site. To hot-drain a filter, puncture the dome end and drain it overnight (12 hours). Oil should be drained when the air temperature is 60°F or higher.

After hot-draining the filter, place it in a sealable plastic bag, coffee can with lid, or other leakproof container. If your community does not collect household hazardous wastes, you might call a local service station, "quick lube," or auto parts retailer. They may accept filters for free or for a modest charge.

Antifreeze

Antifreeze is poisonous. It also has a sweet taste, which can attract pets and small children. It takes only 3 ounces of pure antifreeze to kill an average-sized adult, so take extreme care with this product.

Because of the personal and environmental risks, do not pour antifreeze out in the backyard, at the curb, or into a storm sewer, stream, lake, or river. Do not put it into your garbage for collection, and do not mix it with used oil.

The best option for antifreeze is to transfer it to a sturdy container, clearly mark the contents, and take it to a household hazardous-waste collection program. You might also contact a local service station or an automotive or radiator repair shop to find out if they can dispose of or recycle the waste.

Never dump antifreeze down a household drain or toilet unless you have checked with your sewage treatment plant to find out if it is acceptable and what procedures to follow. Some treatment plants cannot handle antifreeze. If you are told it is safe to dispose of antifreeze this way, make sure you rinse the drain or toilet afterward so pets are not attracted to residual amounts of antifreeze.

Also, never dump antifreeze down a drain or toilet if your home is connected to a septic system, if you have a large amount of antifreeze (more than 1 or 2 gallons), or if your car has a copper radiator (check with your dealer or mechanic). A copper radiator can contaminate antifreeze with lead from the solder used to manufacture or repair the radiator.

A new antifreeze on the market uses propylene glycol, which toxicology studies show to be three to four times *less toxic* than the ethylene glycol traditionally used in antifreeze. But it is not entirely nontoxic, so take precautions with it. Do not leave pans of antifreeze lying around the garage for pets to sample. Follow the same disposal procedures that you would observe with traditional antifreeze.

Lead-acid batteries

Lead-acid batteries are the batteries used in cars, motorcycles, snowmobiles, and other vehicles. Because of the amount of toxic metals and corrosive lead-contaminated acids in these batteries, many states have made it illegal to put them in the trash.

Recycle your batteries! Most stores that sell lead-acid batteries will accept used batteries for recycling. In some states, they are required by law to take old batteries. When storing a battery before recycling it, keep it dry. Keep it in a garage or storage shed, where it is out of reach of children and pets.

Tires

Tires take up a lot of space in landfills, attract insects (such as mosquitoes), and can result in devastating fires. At a Virginia site in 1984, for instance, from 5 to 7 million tires burned for nine months, sending clouds of pollution into the air. In some states, whole scrap tires are no longer allowed in landfills.

The most common way to get rid of old tires is to turn them in to the retailer when purchasing new ones. Most retailers will then make sure the tires are recycled or remanufactured into retreads.

If you have an old tire stashed in the back of the garage—one that was never turned in when new tires were purchased—contact your local or regional recycling coordinator to find out where to take it.

Prepared with Dan Kraybill, Illinois Hazardous Waste Research and Information Center.

Dana Duxbury

Keeping an eye on hazardous waste

- In Washington state, the "Wastemobile" travels to twenty-four sites for two weeks each year, collecting household hazardous waste.

- In Virginia, the "Million Points of Blight" project encourages communities to paint warnings on storm drains to teach the dangers of dumping hazardous products into the system.

- In Lincoln, Nebraska, the household hazardous-waste program hires bounty hunters; it pays residents for turning in any one of five banned pesticides.

Keeping track of these and hundreds of other household hazardous-waste programs is one of the jobs of the Waste Watch Center, a nonprofit organization based in Andover, Massachusetts. With a staff of five and a network of contacts in every state, the Waste Watch Center has data on the nearly 8,000 household hazardous-waste programs conducted across the United States since 1980.

To say the least, the increase in programs has been dramatic. In 1980, there were only two household hazardous-waste programs in the country; but in 1994, there were 1,543, says Dana Duxbury, president of the Waste Watch Center.

The Waste Watch Center also serves as a clearinghouse for information on hazardous-waste issues, and it publishes a quarterly newsletter that follows progress in every state. Through such efforts, Duxbury says the center tries to answer three key questions:

- *What are the products you need to be concerned about?*

- *Why should you be concerned about them?*

- *What can you do about them?*

According to Duxbury, household hazardous waste generally falls into five categories—paint, pesticides, household cleaners, automotive products, and the catchall "other" category (which includes everything from batteries and pool chemicals to glues and explosives).

When these products are improperly discarded, she notes, they can put a strain on a city's wastewater system, pollute streams and lakes, and contaminate septic systems. And since a lot of these products create fumes, they can even create indoor air-quality problems when used inside.

Most of the household hazardous-waste programs across the country hold short-term events, in which people can take their hazardous materials to a site on one particular day or weekend. But there is also a growing trend toward *permanent* drop-off sites. In 1984, there were only seven permanent hazardous-waste drop-off sites in the country. By 1994, there were 227 permanent sites in 31 different states.

In the future, Duxbury sees a continuing expansion in the types of hazardous waste that these sites will accept. For example, some programs now encourage people to bring in their old fluorescent light bulbs—the second leading source of mercury in the waste stream. Batteries are currently the leading source of mercury, but with new designs by manufacturers, that could soon change, Duxbury says.

"There has also been a greater opportunity to recycle paint," she adds. Some paint manufacturers purchase the paint turned in at hazardous-waste collection sites, then process it and sell it commercially. According to Duxbury, the quality of recycled paint is roughly equivalent to that of medium-grade paint.

Other new trends have been an increasing use of mobile units to collect hazardous waste, as well as a focus on redistributing collected materials. For example, if a product is of low toxicity, is in its original container, and meets other criteria, some programs let people come in and pick products off the shelf; other programs distribute the products to nonprofit organizations and businesses.

"Programs are learning how to do things cheaper and wiser," says Duxbury.

25 Dispose of paints and solvents safely

Paint

Most paints contain solvents and metals that are hazardous to the environment. However, an exception to this is the new solvent-free paint (see accompanying sidebar). Latex and oil-based paints should not go to the landfill, nor should they be dumped in storm sewers, household drains, or on the ground.

Here are some disposal options:

Use it up. Try to buy only the amount of paint that you need, and then use it up. Oil-based paint can stay in good shape for up to 15 years, whereas latex paint can last for about 10 years. The paint is probably still good if the paint is labeled, fills about a third of the container, hasn't been frozen and thawed repeatedly, and mixes when stirred. If latex paint has been frozen, brush it on some newspaper. If there are no lumps, you can use it.

Donate it to others. If you are unable to use up your paint, donate it to friends, relatives, churches, recreation departments, community service organizations, or theatrical groups.

Recycle. In some areas, recycling programs will accept paint.

Take it to a household hazardous-waste collection site. If your community does not have a household hazardous-waste site, contact your local or state environmental protection agency for disposal ideas.

The last resort: dry out latex paint. If you cannot use, donate, or recycle latex paint—and if you do not have a local hazardous-waste site—you can dry out the paint and toss it in the trash. Do not dry out oil-based paint because the fumes are hazardous. Also, do not put oil-based paint in the trash.

Dry out latex paint in a well-ventilated area away from children, pets, and rain. It may take days or even months for the paint to dry, depending on the type and amount you are dealing with. The best way to dry out latex paint depends on whether the can contains small or large quantities of paint.

If the can contains small amounts of latex paint:

Step 1. Remove the lid and stir the paint to speed drying.

Step 2. Allow the leftover paint in the bottom of the can to dry out. Periodically stir the paint.

If the can contains large quantities of latex paint:
- Brush paint in layers on newspaper or cardboard.
- *Or,* pour 1-inch layers of paint into a cardboard box lined with plastic. Allow the paint to dry one layer at a time.
- *Or,* mix the paint with cat litter, sawdust, or sand in a cardboard box lined with plastic. Let it dry.

If latex paint has separated:

Step 1. Pour the clear liquid on top into a cardboard box lined with plastic.

Step 2. Mix the liquid with an equal amount of cat litter or other absorbent material.

Step 3. Allow the leftover paint in the bottom of the can to dry out.

Paint thinner and other solvents

Solvents are products used to dissolve other substances. They can be used to thin paint, clean paintbrushes, remove nail polish, clean machinery, remove grease stains, and strip paint and varnish. Many solvents are poisonous and flammable, and they pose hazards to groundwater and surface water.

Use it up. This is always the preferable option.

Reuse solvent that has been dirtied with paint. There are ways you can reuse the solvent used to clean paintbrushes. Let used turpentine or brush cleaners sit in a closed container until the paint particles settle out. (This could take weeks or even months.) Next, pour off the clear liquid, which can be reused. Add an absorbent such as cat litter to the remaining residue, and let it dry completely. Before you do this, however, make sure you know where you can dispose of the dried residue. For ideas, contact your household garbage collection service, a household hazardous-waste program, or a local or state environmental control agency.

Prepared with Dan Kraybill, Illinois Hazardous Waste Research and Information Center.

SOLVENT-FREE PAINTS STIR UP CHANGES

In the wake of new regulations in many states, paint manufacturers are changing their recipes for oil-based paints. Manufacturers are now able to produce oil-based paints that contain lower levels of "volatile organic compounds," which come from solvent chemicals. When volatile organic compounds, or VOCs, are emitted into the atmosphere, they add to the formation of smog.

Latex paints also contain solvents, but not nearly as much as oil-based paints. Nevertheless, paint manufacturers anticipate possible regulations on solvent levels in latex paint as well. That is why one manufacturer—Glidden—came out with a solvent-free latex paint in 1993.

For those who use oil-based paint, the solvent-free variations will call for some adjustments. The National Paint and Coatings Association says solvent-free oil-based paint does not apply as easily as traditional oil-based paint. It also is not as hard, not as resistant to chalking and mildew, and more likely to penetrate porous surfaces.

However, there are also advantages. Solvent-free paint is less likely to crack and peel, emits less odor, and is more fire-resistant. And, of course, it is less of an environmental hazard.

If you would like to try solvent-free paint, either oil-based or latex, you will need to specifically ask for it at the store.

Prepared with Nicholas Smith-Sebasto, environmental education specialist, UI Cooperative Extension Service.

26 Dispose of adhesives, aerosols, household cleaners, and other hazardous waste safely

Adhesives

Adhesives are products that hold things together or fill cracks. They include:

Auto-body filler	Linoleum pastes
Carpet adhesives	Patching pastes
Caulking compounds	Rubber cement
Epoxy resins	Spackling compounds
Glazing compounds	Tile adhesive
Glues	Tile grout
Joint fillers	Wood putty

These products contain solvents and other toxic chemicals. Among glues, the safest choices are white glue, glue sticks, library paste, and yellow glue. Select them whenever possible.

Use it up or donate it to others. If you cannot use up the product yourself, give it to someone who can.

Take it to a household hazardous-waste collection site. Find out where the nearest collection site is located.

The last resort: dry the adhesive. If you cannot use up or donate your leftover adhesive—and if there is no local hazardous-waste site—follow this procedure:

Step 1. Find a well-ventilated area away from children, pets, and sources of heat or flames—preferably outside, especially for amounts more than a few ounces. Wear chemical-resistant gloves, and avoid inhaling fumes.

Step 2. Harden the adhesive. If you have a small amount of adhesive, open the container and let it dry. If the adhesive is in a tube, slit the tube for drying. For larger amounts of adhesive, spread the adhesive in thin layers on cardboard or newspapers. For epoxy, auto-body filler, and other *two-part* adhesives, mix the two parts together before letting them dry.

Step 3. Throw the adhesive away. When the adhesive is hardened and the newspaper or cardboard dries, you can safely place them in the trash.

Aerosol containers

Aerosol containers are pressurized products that sometimes contain flammable or poisonous chemicals. If you dispose of these pressurized containers in the trash, they can be punctured and explode. They can start a fire or injure sanitation workers.

Use it up or give it to others. A can is empty and safe for disposal if you no longer hear air being released from the container. If you cannot empty the aerosol cans by using the contents yourself, find someone who can.

Empty and depressurize the aerosol container. The following procedure explains how to depressurize the container. *Do not use this procedure with pesticides in aerosol cans.*

Step 1. Work in a well-ventilated area away from children, pets, and sources of ignition. Follow safety precautions on the container, and avoid inhaling vapors.

Step 2. Empty the container. Discharge the contents of the container into a cardboard box until you no longer hear the air being released from the can. If the nozzle is plugged, try replacing it with a nozzle from another container. If you cannot empty and depressurize the container, save it for disposal in a household hazardous-waste collection.

Step 3. When the container is empty and depressurized, put the cardboard box in the trash. Empty containers can be recycled.

Household cleaners, medications, and cosmetics

Household cleaners pose a pollution problem if they contain solvents. Your cleaner contains a solvent if the label contains one of these warnings:

- Flammable
- Combustible
- Caution
- Warning
- Danger
- Contains petroleum distillates or aromatic hydrocarbons

If a household cleaner contains a solvent, do not dump it down the drain or put it in the trash. Check to find out if someone else can use it up—as long as the cleaner is in its original container and is properly labeled. You might give it to a family member, friends, churches, recreation programs, or community service organizations.

Household cleaners *without* solvents can be poured down a drain or flushed down a toilet, as can medications and cosmetics. Sanitary sewers and septic systems usually can break down the chemicals in these products. If cleaners are solids, they can also be put in the trash.

A good rule to follow is that if a product goes down the drain during normal use, usually it can be disposed of by pouring down the drain. However, dilute the product with plenty of water during disposal. Also, do not pour more than one product down the drain at a time because some substances may react when they are mixed.

Old appliances

Some appliances, such as refrigerators, freezers, washing machines, air conditioners, microwave ovens, and central heating and cooling units, contain toxic PCBs. Some also contain CFCs, which can cause problems when released into the atmosphere. CFCs are suspected of contributing to the depletion of the planet's protective ozone layer.

Turn in the old appliance when you buy a new one. This is the most common way to get rid of an old appliance. Many retailers will send old appliances to scrap metal processors to recover the metal.

Contact your waste hauler or your local or regional recycling coordinator. If you have an old appliance to get rid of—but you're not planning to buy a new one—contact your waste hauler or your recycling coordinator. Even if they are not equipped to handle old appliances themselves, they may stockpile them for shipment to a scrap steel processor.

Prepared with Dan Kraybill, Illinois Hazardous Waste Research and Information Center.

SORTING OUT THE BATTERY DILEMMA

Where batteries go when they die

When batteries stop going and going, they can encounter many fates in the environment, with these being the most common:

Disposal in a regular landfill or community incinerator. When you toss batteries in the garbage, they may end up in either a community incinerator or an ordinary landfill. The primary environmental risk in battery disposal comes from incineration. Most batteries contain heavy metals such as zinc, cadmium, lead, lithium, manganese, and mercury. Therefore, consumers could be exposed to heavy metals that escape through the smokestack or in the ash residue.

Disposal in a hazardous-waste landfill. If batteries are collected during a community hazardous-waste collection day, contaminants are removed and disposed of in a specially designed landfill.

Recycling. If batteries contain an element of value, such as silver, some manufacturers, retailers, or communities will collect the batteries for recycling. The element is then removed and sold or used for other purposes.

The following provides a brief description of the most common types of batteries and the disposal options for each one.

Carbon-zinc batteries

- Cheapest of the typical cylindrical household batteries.
- Do not last as long as alkaline batteries.
- Contain zinc and cadmium. Virtually all U.S. battery manufacturers have stopped adding mercury to carbon-zinc batteries.
- Nonrechargeable. Not widely recycled in the United States. Throw them away with the trash, or take to a hazardous-waste collection site.

Alkaline batteries

- Can sit on the shelf for a long time without losing their charge.
- Pack more power than carbon-zinc batteries.
- Contain cadmium and zinc. Virtually all U.S. battery manufacturers have stopped adding mercury to alkaline batteries.
- Only newer alkaline batteries, specifically designed for reuse, can be recharged. Not widely recycled in the United States. Throw them away with the trash, or take to a hazardous-waste collection site.

Nickel-cadmium batteries, or ni-cads

- Most common type of rechargeable batteries.
- Contain nickel and cadmium.
- Often come embedded in household appliances; when the appliance is thrown out, batteries go with it. In response to legislation in several states, manufacturers have begun designing products in which rechargeable batteries can be easily removed.
- Are recycled. A trade association is working to develop a national recycling program with collection to be handled by retailers, manufacturers, and municipalities.

Mercuric oxide batteries

- Small, round, flat batteries often used in hearing aids, watches, and calculators.
- Mercury and zinc are the primary elements of concern.
- Small size of batteries calls for vigilance to keep them out of reach of children. Children can swallow them.
- Nonrechargeable. Are recycled.

Silver oxide batteries

- Often used in place of mercuric oxide batteries and give longer service. Contain significantly less mercury than mercuric oxide batteries.
- Nonrechargeable. Are recycled.

Zinc-air batteries

- Relatively new. Used most often in hearing aids, pagers, beepers, and medical devices.
- Mercury and zinc are the primary elements of concern.
- Nonrechargeable. Not widely recycled in the United States. Throw them away with the trash, or take to a hazardous-waste collection site.

Lithium batteries

- Used in cameras, calculators, watches, and computers.
- Have the longest shelf life of any battery.
- Nonrechargeable. Not widely recycled in the United States. Throw them away with the trash, or take to a hazardous-waste collection site.

Prepared with Brenda Cude, consumer economics specialist, UI Cooperative Extension Service.

Managing yard waste, food waste, and other household waste

A model home in Maryland illustrates that a lot of our waste needn't go to waste. The home, constructed by the National Association of Home Builders, includes steel framing made from old cars, insulation made from recovered newsprint, hardwood siding that contains sawmill residues, roofing panels that include recycled computer housings, and a wood substitute made from sawdust and recycled plastic grocery bags.

This may not be everybody's dream house, but it shows that there are innumerable ways to ease the strain on our overflowing landfills. The following section offers many practical and effective alternatives to the ordinary disposal of household waste and yard waste.

27 Reduce household waste

Cutting back

In the United States, each person generates an average of 4 pounds of garbage per day; after recycling, the average amount is still a whopping 3.5 pounds. But with a little thought and effort, you and your household can become—indeed, may already be—involved in the solution to the ever-growing problem of solid-waste disposal.

The best way to manage solid waste is to not produce it in the first place. Here are some key ideas on how to reduce the amount of garbage that you generate:

- Buy only what you need.
- Buy durable products, and maintain and repair them to ensure longer product life.
- Borrow or rent items you don't use often.
- Limit junk mail. When obtaining credit cards or new magazine subscriptions, ask that your address not be given out to mailing lists. Also, write to the Direct Marketing Association at the following address and ask that your name be eliminated from mailing lists:

Direct Marketing Association
Mail Preference Service
P.O. Box 9008
Farmingdale, NY 11735-9008

To have your name eliminated from additional lists, you can write to the three major credit-reporting agencies. Request that your name be removed from their direct-marketing files. The addresses are:

Trans Union, Inc.
Name Removal Option
P.O. Box 97328
Jackson, MS 39288
(800)241-2858

TRW-NCAC
Target Marketing Services Division
701 TRW Parkway
Allen, TX 75002
(800)353-0809

Equifax Options
Equifax Marketing Decisions Systems
P.O. Box 740256
Atlanta, GA 30374-0123
(800)880-1184

That's a wrap-up

Reducing the amount of packaging you buy is another important way you can cut back on waste. Packaging accounts for 30 percent of the total volume of municipal waste and is often cited as a major reason why many landfills are nearing capacity. In addition, packaging costs money. It is estimated that consumers spend $1 out of every $11 for packaging.

As you think of ways to reduce packaging, it helps to understand when it is sometimes necessary. Packaging provides detailed information, protects products from harm, provides a convenient way to carry and store products, provides convenient ways to use a product (microwaveable packaging and single-serving containers, for example), protects the consumer with tamper-evident and child-resistant systems, enhances the image of a product, and reduces food spoilage.

Despite these purposes, many products are overpackaged in multiple layers and packages within packages. So it's important to evaluate what's out there.

Evaluating packaging

To avoid excess packaging, compare the size of the package to the size of the product. If there appears to be far more package than product, choose another brand.

You can also avoid excess packaging by choosing large containers instead of several small ones, buying concentrates, and buying food in bulk when it is available, instead of buying it prepackaged.

To help consumers evaluate packaging, the Coalition of Northeastern Governors has developed the following "preferred packaging guidelines":

No packaging. Some products don't need a package. For instance, is it really necessary to shrink-wrap produce and place it on a paperboard tray? A hammer could be hung on a rack, rather than put in a blister pack. However, other products clearly do require some packaging.

Minimal packaging. "Minimal packaging" is when a package performs its function with a minimum amount of material. For instance, a manufacturer can do this by reducing the weight of a package. In 1970, a plastic milk jug weighed 95 grams, but today it weighs only 60 grams. Aluminum cans were 26 percent lighter in 1989 than in 1972.

Manufacturers can also redesign a product to minimize packaging; they can reduce multiple layers of packaging to a single layer. One example would be a bottle with no outside packaging, rather than a bottle in a cardboard box.

Notice that the products with greater convenience often use the most packaging. Individually packaged juice, raisins, pudding, vegetables, and other food items are clearly more convenient, but they greatly increase the amount of packaging.

Returnable packaging. Many containers, such as glass milk bottles, can be returned to the industry for redistribution and reuse.

Refillable packaging. Refills are now available in many stores for an increasing number of cleaning products. Also, many hair salons offer refills on shampoo and other hair-care products.

Reusable packaging. Reusable packaging is anything that can be reused for a different purpose; for example, a plastic container can be reused as a storage container. However, never reuse food packages for storing hazardous materials. Reuse food containers only to store the same type of food that was originally in the package.

Recyclable packaging and packaging made from recycled materials. "Recyclable" and "recycled" mean two different things. A recycled package is one that has been made from recycled materials; a recyclable package is one that *can be* recycled.

Chapter 28 provides tips on how to identify both recyclable and recycled products.

Prepared with Brenda Cude, consumer economics specialist, UI Cooperative Extension Service.

28 Recycle paper, glass, and aluminum (and a few other household wastes)

Cliché but true

Textile scraps turn into shoes. Old tires turn into gasoline. Glass turns into pavement. And old paint turns into new paint. These are just some of the ways that our garbage is being resuscitated—brought back to life through recycling.

It may be a cliché, but you can make a ton of difference by simply recycling a few common household waste products.

The U.S. Environmental Protection Agency estimates that commercial and residential rubbish in the United States amounts to about 207 million tons a year. By weight, about 37 percent of all municipal solid waste in the United States is paper and paperboard, followed by yard trimmings, plastic, metal, food waste, glass, and wood.

Much of this material can be recycled, which keeps it out of the landfill and reduces the amount of virgin material used to make products.

The material that your local recycling program will accept depends on the program's ability to sell the material for a reasonable price. As markets change, a local program may add or delete a certain type of material from collection. The instructions that local recycling programs offer to

Total waste generated each year in the United States
207 million tons

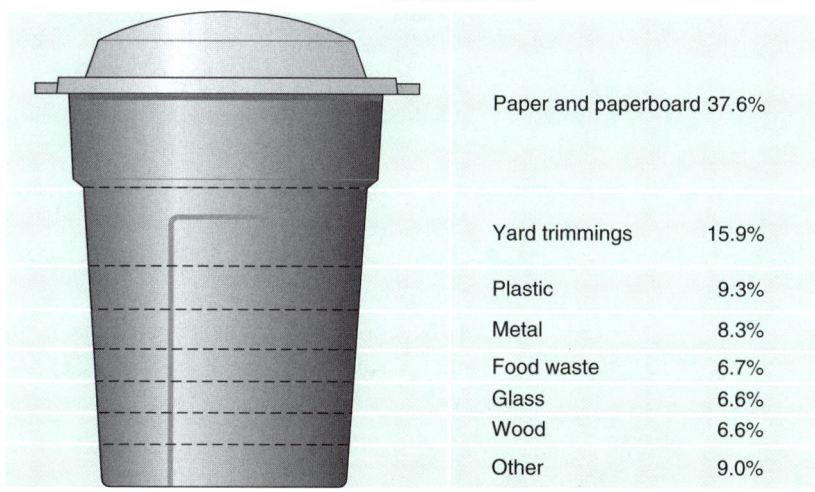

Paper and paperboard	37.6%
Yard trimmings	15.9%
Plastic	9.3%
Metal	8.3%
Food waste	6.7%
Glass	6.6%
Wood	6.6%
Other	9.0%

Source: U.S. EPA. *Characterization of Municipal Solid Waste in the United States: 1994 Update*, 1994.

customers on how to prepare materials for collection are usually designed to increase the market value of materials. Different recycling programs have different guidelines, so be sure to check locally for specifics.

Paper

Here are some tips for recycling paper:

- Separate junk mail, telephone books, and magazines from other types of paper. Mixing paper that contains contaminants (such as glues) with other paper could mean that the entire load must go to a landfill instead of being recycled.
- Keep the paper clean and dry.
- Flatten and bundle corrugated cardboard, which consists of two layers of heavy cardboard with a ribbed section between them.
- When recycling high-grade papers, including typing, notebook, photocopy, and writing paper, separate white paper from colored paper, and box or bag each separately.
- Many collection programs do not accept carbon paper and papers with residues; cellophane windows; self-stick adhesives; or wax, plastic, or foil coatings.

Metals

Because metals account for the second largest percentage of our rubbish, cans probably should be next on your list of household wastes to be recycled. Here are some guidelines:

- More than 90 percent of all beer and soft drink cans are made of aluminum. To recycle, rinse them and box or bag. Crushing is not necessary, but it saves space.
- Aluminum foil, pie pans, TV dinner trays, and lawn furniture are collected for recycling in some communities.
- Beverage containers in which only the tops or bottoms are aluminum may be recyclable, but should not be mixed with pure aluminum.
- Food cans usually are called "tin cans" but are actually made of steel or tin-coated metal. They constitute 37 percent of total can production. They can be recycled, so rinse them and remove the label to prepare them for pickup.

Glass

All kinds of glass containers—heavy or light—can be recycled. In addition, glass can be recycled and reused an indefinite number of times.

The following are some important points to keep in mind:
- Clear, green, and brown glass are collected in many municipal recycling programs.
- Paper labels can be left on glass to be recycled, but aluminum neck rings and caps can be a problem, depending on the recycling equipment being used.
- Recycling centers will not accept light bulbs, ceramic glass, dishes, or plate glass because these items consist of different materials than bottles and jars.
- Cullet, or crushed glass, can be used to make new bottles, jars, and other containers. Some other uses for cullet are "glassphalt" (a road paving material), building panels, and terrazzo. Although resale value is lower than some other recyclables, markets are relatively stable.

Food and yard waste

Because food and yard waste accounts for nearly 25 percent of our solid waste, dealing with it some other way makes good sense. Many, but not all, kitchen scraps make good compost. For information on what food waste can go into a compost, as well as details on composting yard waste, see Chapter 31.

Plastics

The markets for recycled plastic have been slow to develop, but the number of new products made from recycled plastic is growing. Plastic soda and milk bottles can be shredded and made into lumber substitutes, packaging, fabrics, chemicals, machine parts, and household items such as pans, flower pots, and fiberfill used in sleeping bags, vests, and carpeting.

Although there are many types of plastics, only two are currently being recycled in any significant quantities: PET (polyethylene terephthalate), the primary plastic for soda bottles; and HDPE (high-density polyethylene), the usual component of milk jugs and detergent bottles. Mixing different types of plastics can increase problems in the recycling process, so recycling centers use a coding system to separate them. Even containers made from the same plastic may not be mixed together if they were formed using different processes. To determine which kind of plastic a product is made of, look for the plastic container code (see the accompanying chart).

If your local recycling facility accepts plastics, find out which kinds and whether paper labels can be left intact. Rinse plastic containers, remove caps, and flatten them so they take up less space.

Code system for plastic containers

Code	Material	Typical use
♳ PETE	Polyethylene terephthalate (PET)	Soft drink bottles
♴ HDPE	High-density polyethylene	Milk jugs Laundry detergents
♵ V	Vinyl/polyvinyl chloride (PVC)	Vegetable oil bottles
♶ LDPE	Low-density polyethylene	Dry-cleaning bags Bread bags
♷ PP	Polypropylene	Yogurt cups
♸ PS	Polystyrene	Carryout containers
♹ Other	All other resins plus materials made with layers of different plastics	Microwaveable serving ware

Many plastic products display one of the above codes to indicate the kind of plastic from which it was made. PET and HDPE are the most commonly recycled plastics. Find out which ones your recycling center will accept.

Motor oil

Motor oil never wears out; it only gets dirty. To recycle it, drain the car, motorcycle, or lawn mower oil into a container with a sealable lid. Some garages, service stations, and large retailers with auto shops accept used oil. You can also contact your municipal department of public works to locate a local recycling station. Once impurities are removed, used oil can be marketed as re-refined oil or industrial fuel oil.

> ✓ **CHECK IT OUT**
> For more information on disposing of automotive products: page 123.

Look for recycled products

So much hinges on the market. If a recycling center does not have markets where it can sell collected material, it cannot cover its costs. Therefore, help sustain the markets that *do* exist by purchasing products that have been produced from recycled material.

Look for the "recycled" symbol—three arrows forming a circle or triangle. However, note the similarity between the "recycled" symbol and the "recyclable" symbol (see illustration). Even though the symbols look almost identical, they mean different things. When a product is labeled "recycled," it has been produced—at least in part—from recycled materials. When it says "recyclable," that means it *can* be recycled.

To help prevent confusion between the two terms, a statement usually says whether the product or packaging uses recycled materials.

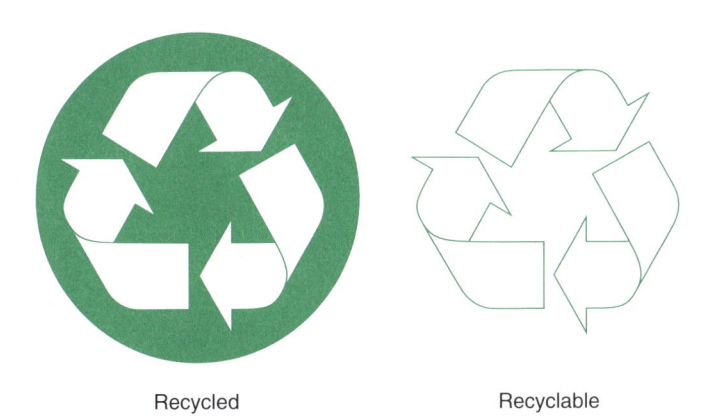

Recycled   Recyclable

The "recycled" and "recyclable" symbols may look alike, but they mean different things.

Many products sold in paperboard boxes, such as cereals, cookies, detergents, baking goods, and snack crackers, are packaged in containers that use recycled newsprint. If the paper has been recycled, the interior of the box will often be gray rather than white.

In addition, most glass and aluminum containers include some recycled materials, although these containers rarely display the recycled symbol. You will also find several paper products that are made from recycled paper, including bathroom tissue, greeting cards, paper towels, writing paper, and envelopes.

Many labels now specify the percent of "postconsumer waste" that went into the product. Postconsumer waste is the term for material that businesses and consumers supply to recycling programs, while "preconsumer" waste is generated by manufacturers and industry. Buying products that contain "preconsumer" waste also supports recycling.

Keep in mind that when labels use the term "recycled," the percentage of recycled material may be very low or the product and packaging may be made *entirely* from recycled materials. Also, just because a package is labeled "recyclable," that doesn't always mean you can recycle it in your local community.

Prepared with Brenda Cude, consumer economics specialist, and Kathleen Brown, community leadership and volunteerism educator, UI Cooperative Extension Service.

Additional source: Reduce-Reuse-Recycle: Alternatives for Waste Management, *by Marie Hammer, professor of home environment–solid waste management, Florida Cooperative Extension Service.*

KEEPING CONTAMINANTS OUT OF THE RECYCLING STREAM

What you include in your recycling load makes a difference. When certain materials get into the recycling system, they can disrupt equipment or downgrade the final recycled product. A low-grade product, in turn, makes it harder for a recycling center to sell the recycled materials for reuse.

Therefore, make sure you know what materials are accepted by your local recycling center. The following items are considered "contaminants" in the recycling stream by some reprocessors and manufacturers and should not be included among your recyclable materials. However, check with your local recycling centers or processors. Some state-of-the-art reprocessing facilities can handle certain contaminants, but others may not be able to do so. Policies vary on what is considered a contaminant.

What constitutes a contaminant?

Materials being recycled	Common possible contaminants	Possible effects
Paper	• Food residue and wrappers/ packages contaminated by food • Laminated or wax paper • Blister packs (and multi-materials packaging) • Foam and plastic wraps, films, and tapes • Plastics, glass, and metals	Food and drink containers attract pests. Plastics, waxes, and glues may clog some machinery and affect the color, finish, and ink-holding quality of recycled paper. Cereal boxes and decorative packaging may add a high ink content, often have plastic liners or components, and may downgrade the final recycled product (unless separated out).
Cans	• Soda in cans • Food residues	Food and soda attract insects and vermin, cause spills, and create odors.
Glass	• Lids • High-temperature glass • Ceramics • Mixed colors • Metals	Metal lids may jam some glass-crushing machinery and affect the end product when the glass is melted for reuse. The end product is also affected by light bulbs; lab glass; window glass; windshields; ceramics; and glass that is resistant to heat, chemicals, and electricity (such as Pyrex).

What constitutes a contaminant?, cont.

Materials being recycled	Common possible contaminants	Possible effects
Plastics	• Lids and other closures • Glass and some foreign materials • Other plastic resins	Closures may prevent containers from being compacted for economical collection and shipment. Also, closures are often made from a different material than the container itself. The composition of plastics varies. Therefore, some types cannot be blended together and must be sorted according to the codes on the item.

Prepared with Shirley Niemeyer, home environment specialist, University of Nebraska–Lincoln Cooperative Extension Service; and Marie Hammer, professor of home environment–solid waste management, University of Florida Cooperative Extension Service.

Illinois

Small recyclers work together to make trash pay

There's more to recycling than collecting paper, glass, metal, and plastic. Recyclers also need to find manufacturers that will buy the material for a good price. But this can be a tough chore, especially for small, rural recycling operations, says Pam Walkenbach, a specialist in recycling market development for the University of Illinois Cooperative Extension Service.

Employees sort paper at the Community Recycling Center in Champaign-Urbana, a member of the Recycling Cooperative of Illinois.

In 1994, for example, some of the small recyclers in Illinois were unable to find markets where they could sell the cardboard they had collected. So they gave away the cardboard to larger recyclers, who turned around and found a manufacturer that bought the cardboard for a substantial profit.

Small recyclers do not always have the resources or the time to find good markets for the material they collect, and they often do not generate the volume of material that attracts buyers, Walkenbach says. To remedy this problem, the Recycling Cooperative of Illinois (RCI) took shape in 1994 as a way for recyclers to help each other find good markets for their product.

It is one of the first cooperatives in the United States to serve all types of recycling operations, both private and government-run, says Walkenbach, co-chair of the RCI steering committee. Most other cooperatives have been set up to serve primarily government programs.

The cooperative's full-time executive director helps to locate buyers and good prices for the material collected. In addition, recyclers in the cooperative ship their material together; that way, they can offer a large enough volume to attract buyers. And by sharing in the cost to truck the material, they cut transportation expenses.

Another benefit is that member recyclers cut costs by making bulk purchases of products and supplies—baling wire, equipment, and curbside containers for recycled material.

The cooperative's marketing assistance will especially help with tough-to-sell material, such as plastics. Many recyclers lose money on plastics, Walkenbach says, because they are forced to unload them wherever they can.

To make matters more difficult, plastic is easily contaminated, and its light weight means less money per load (buyers pay by the pound). In contrast, metals are easier to handle because they have been recycled for the longest time and the markets are well developed.

As the Recycling Cooperative of Illinois moves through its first year, Walkenbach says the key will be seeing how well the members—who are competitors—can work together.

The cooperative started with thirteen recyclers, and there is a lot of potential for growth, Walkenbach notes. Illinois has more than two hundred curbside collection programs and another three hundred drop-off recycling sites in operation.

29 Reuse and respond

Reuse

The Tri-County Household Hazardous Waste Program in St. Cloud, Minnesota, is one of a growing number of facilities that offer an innovative service. After inspecting products that come to the collection site, operators decide which ones qualify for a product exchange program. Products that pass inspection are made available for people to come in, pick them up for free, and reuse them. The program targets three main markets: homeowners and businesses; social service agencies; and commercial users of large volumes.

As you can see, finding ways to reuse products takes both common sense and creativity. But such creativity is not limited to community organizations. Here are a few ideas on how you can get double duty out of various products around the home:

- If you can't reuse a product yourself, find someone who can. Sell it in a yard sale, donate it to a community organization, or set up an exchange program.
- Consider donating books and magazines to public libraries or to nursing homes.
- If newspaper recycling is not an option in your area, check with local pet stores and animal shelters to see if they can use your discarded newspapers.
- Reduce the number of paper bags your family uses by carrying reusable tote bags to grocery and department stores. Attractive nylon mesh bags, available from many sources, can be stored easily in the glove compartment of your car. Another option is to use durable canvas bags. Also, reuse the bags that you do receive.
- Minimize use of paper towels, paper plates, and napkins. Invest in cloth napkins for everyday use, and choose reusable cloths, towels, and plates, rather than paper "throwaways."
- Call local elementary schools and day-care centers to find out what items might be useful; local teachers may consider your trash to be treasure in the hands of creative youngsters. Many teachers want aluminum containers, beads, beans, bottles, boxes, brushes, buttons, burlap, old calendars, candle stubs, carpet scraps, used greeting cards, cloth scraps, wallpaper samples, coat hangers, coffee cans, gift wrap remnants, magazines, mirrors, oatmeal boxes, paper bags, pie tins, plastic bags, rug samples, toilet paper rolls, and yogurt containers.
- Substitute rechargeable batteries for throwaway batteries.

- Avoid other single- or limited-use items, such as throwaway cleansing pads and cigarette lighters, nonrefillable ink pens, and foil baking pans.
- Purchase beverages in refillable bottles. If they are not available, ask the merchant to stock them.

Respond

Reduce, recycle, and reuse are known as the three Rs of the waste-reduction movement. But a fourth R often is forgotten: Respond.

Like voting, shopping means making choices. If you find products that you consider unacceptable, don't buy them. You may refuse to buy packaging that you think is excessive, packaging that is not recyclable locally, or products that are harmful to the environment.

You can send an even stronger message to the manufacturer by writing a letter to voice your concerns. Explain why you think the product is not packaged well or other reasons you may have for not purchasing it. Offer any suggestions you have for improvement.

When selecting a store, consider those that make the following good environmental choices:

- Minimize prepackaging of produce.
- Allow customers to reuse store bags or accept bags and other materials for recycling.
- Offer a bulk-buying section for grains, pasta, seeds, and other dry products.
- Offer their own line of products and packaging that uses recycled materials.
- Recycle their own shipping materials and store-generated waste.
- Offer refillables or returnables for some products.
- Use discretion in labeling products with environmental terms such as "green" and "recyclable."

After shopping, be sure to follow through. Tell the management that you value their steps to lessen the impact of products on the environment. And if you bought a product because the packaging was recyclable, then recycle it. If you did not buy a product because of environmental concerns, write to the manufacturer.

You are likely to find that environmental shopping reduces your expenses because less packaging usually means lower cost. Sometimes, however, the environmentally preferred option will cost *more*. Then, you will have to make a choice.

Prepared with Brenda Cude, consumer economics specialist, and Kathleen Brown, community leadership and volunteerism educator, UI Cooperative Extension Service.

ENVIRONMENTAL TRADE-OFFS

Environmental choices are not always as simple as we like to believe. For instance, one of the great debates is whether paper or plastic bags are better. At one level, the answer seems obvious. Paper is a renewable resource that should biodegrade easily. Plastic is made from nonrenewable petroleum products, and most plastic will likely take decades to degrade.

On the other hand, plastic bags take up less space than paper bags in a landfill. A thousand paper grocery bags stack to about 4 feet, whereas the same number of plastic bags stands only 4 inches high. In addition, the amount of sulfur dioxide, nitrogen oxide, carbon monoxide, and dust produced in the manufacture of paper bags is much higher than in the production of plastic bags. Wastewater pollutants are also many times higher during the production of paper bags.

The classic dilemma of cloth versus disposable diapers presents another example of environmental trade-offs.

Cloth diapers versus disposable diapers

Cloth diapers can be reused many times, so they save landfill space	*but*	they may produce more air pollution and water pollution than disposable diapers because of laundry and delivery services.
Disposable diapers generate 100 times as much solid waste as cloth diapers	*but*	growing, harvesting, and producing cotton requires extensive use of chemicals and, in some areas, leads to severe soil erosion.
Cloth diapers cost consumers less and may be more comfortable for the baby	*but*	disposable diapers guard against diaper rash, leak less, and are more convenient.

Prepared with Brenda Cude, consumer economics specialist, and Kathleen Brown, community leadership and volunteerism educator, UI Cooperative Extension Service.

30 Recycle grass clippings

The green waste dilemma

The Illinois Department of Energy and Natural Resources estimates that lawn care for an average midwestern residence generates some 750 pounds of grass clippings a year. Multiply that times the number of lawns in your town, and you can see how quickly the tons of "green" waste mount up.

At the same time, numerous municipalities have begun to refuse to pick up grass clippings and leaves because of reduced landfill availability; in fact, some states have banned yard waste in landfills. As a result, many homeowners are bagging clippings to be collected and sent to a regional composting facility (where one exists), while others have turned to using mulching lawn mowers. A mulching mower reduces the size of clippings and spreads them back on the grass, where they rapidly decompose and return to the soil.

Some homeowners are concerned that leaving clippings on their lawns will contribute significantly to thatch buildup, but research has shown that doing so does not have to cause problems.

Grass clippings are a valuable organic source of nutrients, especially nitrogen. As clippings decompose, these nutrients become available for use by the grass plant. Some mulching-mower users have found that yearly nitrogen applications may be reduced by one-fourth when grass clippings are returned to the turf.

Although mulching mowers and mulching attachments for existing mowers can reduce the clipping size, thus increasing the rate at which grass clippings decompose, you can get the same effect with a normal rotary mower. Mowing on a regular basis with a sharp blade does not reduce the clipping size, but it usually produces clippings that decompose fairly quickly. To find out the recommended mowing heights for different grass species, check the table on page 8.

Thatch management

Of course, it is possible you will encounter problems with thatch, whether you use a mulching mower or not. Thatch is a tightly intertwined layer of dead and living grass stems and roots that can develop between the soil surface and green vegetation. This layer develops when dead organic matter accumulates faster than it decomposes. While some thatch (less than ½ inch) gives resiliency to turf and is considered beneficial, excessive amounts can cause problems.

Thatch can be controlled with vertical mowing, core aerification, and proper watering and fertilization. These techniques, which are described on pages 7 and 8, keep pest problems to a minimum, reducing the need for any pesticides. Also, healthier grass plants can better use any fertilizer nutrients that are applied. And organic materials, such as grass clippings, will decompose faster, releasing their nutrients for use by future grass plants.

Some yards look better

The benefits of leaving grass clippings on the lawn are stressed by programs such as Milwaukee's "Just Say Mow!" Participating cities report reduced garbage loads, and homeowners are discovering healthier lawns, trees, and shrubs. In Fort Worth, Texas, participants in the "Don't Bag It" program said, on average, they felt that their lawns looked 30 percent better when they let the clippings remain than when they previously had collected the clippings.

If you prefer not to leave clippings on the lawn, or if raking up clippings is a good job for the kids, you may want to consider using collected clippings as mulch (a protective covering over otherwise bare soil) or as a component in a compost pile (which in turn will produce nutrient-rich soil conditioners). In either case, you can help ease the landfill overfill and gain a number of free landscape benefits as well.

If you have recently treated your lawn with herbicides, it is a good idea to wait for three mowings before using grass clippings as mulch. A lot depends on the specific herbicide used, but it is possible for herbicide residues in the grass to damage garden crops if you use recently treated clippings as mulch. If you put clippings in a compost, on the other hand, pesticide residues rapidly degrade.

Prepared with Thomas Voigt, turfgrass specialist, UI Cooperative Extension Service.

Additional source: Lawn Care Practices to Reduce the Need for Fertilizers and Pesticides, Fact Sheet AG-FO-5890-B, Minnesota Extension Service, University of Minnesota.

31 Create a compost pile

Making the most out of waste

A compost pile is a teeming microbial farm. The microbes in a compost pile go to work for you, breaking down leaves, grass clippings, fruit and vegetable waste, and other organic matter. When you mix compost material in your yard or garden, you return the organic matter back to the soil, where it offers these benefits:

- Loosens heavy clay soils, allowing better root penetration and improving plant growth
- Improves the infiltration and percolation of water through the soil
- Improves a sandy soil's capacity to hold water and nutrients
- Adds essential nutrients to any soil
- Improves the health of your plants by improving the soil
- Provides a natural, healthy way to dispose of organic-matter wastes
- Creates a healthy environment for earthworms
- Reduces the need to buy trash bags and soil conditioners
- Reduces disposal costs in towns that charge for hauling yard waste

How to make compost

Begin by removing the grass and sod cover from the area where you plan to construct your compost pile. Place a layer of chopped brush or other coarse material on top of the soil surface to allow air circulation around the base of the heap. Then mix together the following components:

- Leaves, grass clippings, or sawdust. These materials should be slightly damp—as moist as a wrung-out sponge, or "sponge damp," as it is called.
- Soil. Adding soil will increase the number of microorganisms to the heap, helping to speed the composting process. However, this is optional because the microorganisms needed for composting are usually present.
- Manure or fertilizer. Adding manure or fertilizer provides the nitrogen needed by the microorganisms. If the manure is dry, add water. On the top of the compost pile, scoop out a "basin" to catch rain water.

The compost process

Once you have created a compost pile, you can typically expect the following changes:

- Two weeks later. A properly made compost heap will reach a temperature of 140° to 160°F in one to two weeks. At this time, you will notice the pile settling—a good sign that the heap is working properly. Initially, the compost pile will be very acidic (a pH of 4.0 to 4.5), but as the process nears completion, the pH will rise to about 7.0 to 7.2.
- Five to six weeks later. At this time, transfer the materials into a new pile. When doing this, turn the outside of the old heap into the center of the new pile. Add water if necessary. You shouldn't need to turn your heap a second time.
- Three to four months later. The compost should be ready to use within three to four months after you first constructed the pile. A heap started in late spring can be ready for use in the autumn. Then you can start another heap in autumn for use in the spring.

You can make compost even faster by turning the pile more often. Check the internal temperature regularly; when it decreases substantially (usually after thirty days), turn the pile.

How can you use compost?

Compost is ready to use when it is dark, brown, and crumbly, with an earthy odor. Let it stabilize for a few extra days, and screen it through a ½-inch screen if you want the finest product. To use the compost, turn your soil, apply a 1- to 3-inch layer of compost, and work it in well.

If you apply compost on a lawn, be sure it is finely ground and that you have a way of working the compost into the soil; otherwise, it could smother the grass. One way to incorporate compost into the soil is to aerate the sod, then apply a ⅛-inch to ¼-inch covering of fine compost. Use a rake to distribute the compost into the crevices.

You can also blend fine-textured compost in potting mixtures. However, make sure the compost does not make up more than one-fourth to one-third of the potting mixture's volume.

What can you put into a compost pile?

Anything growing in your yard is potential food for the micro-organisms in a compost pile. Almost any kind of plant material—including leaves, garden debris, and shrub prunings—will make good composting materials. Chopping your garden wastes with a garden tool or running them through a shredding machine or lawn mower will speed their decomposition.

However, not everything belongs in a compost pile. Here is a handy guide:

Materials that *can* be put in a compost pile

Aquatic weeds	Leaves
Bread	Paper (shredded)
Coffee grounds	Sawdust
Egg shells	Sod
Evergreen needles	Straw
Fruit	Tea leaves
Fruit peels and rinds	Vegetables
Garden wastes	Wood ash
Grass clippings	Wood chips

Materials that should *not* be put in a compost pile

Bones	Fish scraps
Cat manure	Meat
Cheese	Milk
Chicken	Noxious weeds
Dog manure	Oils
Fat	

Adding fat, bones, animal parts, or grease to the pile will cause an odor that will attract rodents, vermin, and the neighborhood dogs. Also, adding noxious weeds to the pile can cause problems if the compost does not become hot enough. The weed seeds may survive; then, when you use the compost material, you may end up putting weeds back into your garden soil.

How large should you make the compost pile?

If you want to produce fast, hot compost, make sure the volume of material in the compost pile is not too small or too large. If the pile is too small, it will have trouble holding its heat. If it is too large, not enough air will reach the microbes in the center of the pile.

As a general rule for homeowners, the compost pile should be between 3 feet and 5 feet high.

Is there an ideal ratio for compost mixtures?

Everything organic has a ratio of carbon to nitrogen in its tissues. A ratio of 30 parts of carbon to 1 part of nitrogen is ideal for the activity of compost microbes. To achieve this balance, mix 2 parts of moist (high-nitrogen) materials with 1 part dry (high-carbon) materials. This combination is the backbone of most compost systems.

If the carbon-to-nitrogen ratio is too high, your compost will not have enough nitrogen, and decomposition will be slower. If the ratio is too low, compost will have too much nitrogen, and odor problems can result.

The following guide provides estimates that can help you judge the ratio of your compost ingredients. As a general rule, blend high-nitrogen materials with high-carbon materials.

High-nitrogen and high-carbon materials

High-nitrogen materials	Average carbon:nitrogen ratios
Soil, typical brown silt loam	12:1
Coffee grounds	20:1
Cow manure	20:1
Grass clippings	20:1
Horse manure	25:1
Fruit, vegetable waste	35:1
Horse manure with litter	30-60:1

High-carbon materials	Average carbon:nitrogen ratios
Cornstalks	60:1
Leaves	60:1
Straw	80:1
Bark	200:1
Sawdust	500:1
Wood	700:1
Paper	800:1

Compost trouble-shooting chart

Symptoms	Problem	Solutions
Compost has a bad odor.	Not enough air.	Turn compost; add dry material if the pile is too wet.
Center of the pile is dry.	Not enough water.	Moisten and turn the pile.
Compost is damp and warm only in the middle of the pile.	Pile is too small.	Collect more material and mix the old ingredients into a new pile.
The heap is damp and sweet-smelling, but still will not heat up.	Lack of nitrogen.	Mix in a nitrogen source like fresh grass clippings, manure, or fertilizer.

Prepared with David Williams and James Schmidt, horticulturists, UI Cooperative Extension Service.

TYPES OF COMPOSTING BINS

Simple bins

One of the simplest ways to create a compost bin is by using a 25-foot roll of 3-foot fencing. Cut off 12½ feet of fencing, form it into a cylinder, and wire the ends together. If necessary, use four 4-foot light-gauge fence posts to stabilize the bin. (Keep in mind that a bin with posts will be more difficult to move.) The materials will make two bins.

You can create another simple bin by drilling three rows of holes, 4 to 6 inches apart, around the sides of a garbage can. Also, drill holes in the bottom of the garbage can. The holes allow air movement and the drainage of excess moisture. Before putting compost in this bin, line the bottom of the can with 2 to 3 inches of dry sawdust, straw, or wood chips to absorb moisture and help the compost drain.

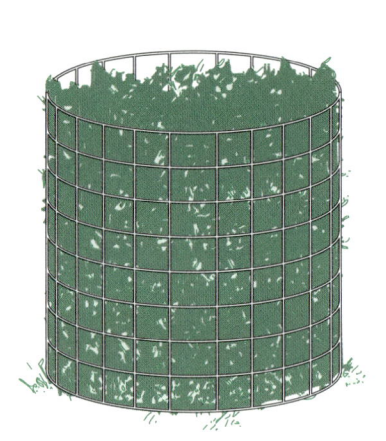

A simple garbage-can bin

A simple wire-mesh bin

Turning bins

A turning bin can come in one of two basic forms. It can be a series of three or more bins, or it can be a mounted, rotating barrel. Either type of turning bin allows you to make compost in a short time by turning the materials on a regular schedule.

If you have a series of three or more bins, turn the compost by moving it from one bin to the next. Turning bins are most appropriate for gardeners with a large volume of yard waste and the desire to make high-quality compost.

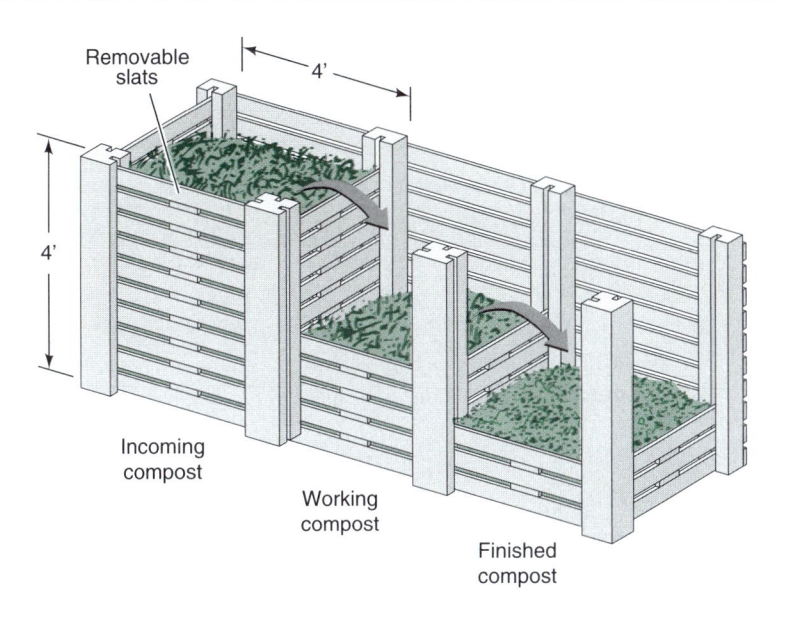

A turning bin makes it easy to turn compost. Simply move the compost from one bin to the next. This speeds decomposition.

Turning with one bin

You can also turn compost with only one bin. Remove the bin from around the pile when it's time to turn the compost. Set up the empty bin nearby and fork the material back into it.

Pit composting

A simple way to compost without a bin is to dig a pit 18 to 24 inches deep in your garden to serve as a composting vessel. The length and width of the composting pit depends upon available space. Add a 1-foot-thick layer of organic matter to the pit, and cover it with a thin layer of soil. This soil covering will reduce odors, provide an inoculum (bacteria) for the composting process, and prevent the pit from attracting rodents or other animal pests.

After the pit is full, cover it with the remaining soil that you originally removed to create the hole. After filling in the hole, the pit will slowly sink. Your garden will be ready for planting when the soil covering the composting pit stops sinking.

Prepared with David Williams and James Schmidt, horticulturists, UI Cooperative Extension Service.

32 Use yard waste as mulch

When a leaf falls in the forest...

Green waste, such as tree limbs, leaves, and grass, are not a problem in the forest. The masses of dead organic material that fall in the woods every year are simply part of a larger nutrient cycle. Homeowners might do well to take the same long view as Mother Nature by putting yard waste to good use as organic mulch. Placing organic mulch around trees, shrubs, and other plantings can benefit your landscape in a variety of ways:

- Conserving soil moisture
- Maintaining uniform soil temperature by insulating the soil
- Minimizing soil erosion and compaction, which can result from heavy rains
- Reducing weed problems by preventing surface germination of weed seeds (however, make sure your mulch is free of weed seeds)
- Giving a neater, more finished appearance to a flowerbed, garden, or other planting
- Altering the soil structure to increase root growth

Applying mulches

Apply mulch about 4 inches deep (certainly no deeper than 6 inches), and keep it at least 6 inches away from tree trunks.

With *established plants* in the garden, apply mulch in mid-spring, when the soil has warmed up sufficiently for active root growth. If you apply mulch before this time, it will keep the ground cool and delay root development.

With *newly planted material,* apply mulch after the plants are set in place and watered well. If you plant in the late summer or early fall, apply the mulch immediately after watering the plants so that soil will be kept warm during the cool nights.

Keep organic mulches wet when applying them to the soil. Extremely dry mulches act as a blotter, removing moisture from the soil.

Some key points

Remember this about mulch:

- Organic mulches have little effect on the nutrient level in the soil and should not be considered a substitute for fertilizer.

- Quickly decaying mulches—such as fresh leaves, wood chips, and straw—can remove a considerable amount of nitrogen from the soil. If you don't add nitrogen fertilizer regularly, a nitrogen deficiency can result.
- If organic mulches are kept too wet, undesirable organisms (fungi, bacteria, and molds) can develop. Also, insects and rodents can overwinter in some organic mulches.
- Although acid sphagnum peat usually lowers the pH, most other organic mulches are slightly alkaline (pH above 7.0).

Yard waste as mulch

Several types of yard waste can make good mulch:

Compost. Compost rates as one of the best organic mulches.

Lawn clippings. Grass clippings are best when used dry. If applied fresh, spread them loosely; otherwise, they will mat down, produce heat during composition, and give off an offensive odor. If the lawn was recently treated with pesticides, wait for three mowings before using grass clippings as mulch.

> **CHECK IT OUT**
> For instructions on how to make compost: page 156.
> For information on recycling grass clippings: page 154.

Leaves. Leaves make a better mulch if they are composted. You can add leaves to a compost that includes various types of yard wastes, or you can compost leaves by themselves to create "leafmold." Compost and leafmold are excellent mulches and soil conditioners.

Wood chips. If you need to dispose of branches and other woody material, one option is to rent a wood chipper. The resulting wood chips are very durable and make an excellent material for covering paths and walkways. If used on landscape beds, however, nitrogen deficiencies can develop if you do not periodically apply fertilizer. You can also obtain wood chips from garden centers, arborists, power companies, and municipal yard-waste facilities.

For information on other organic mulches, see the accompanying sidebar.

Prepared with David Williams, horticulturist, and Sandra Mason, horticulture educator, UI Cooperative Extension Service.

OTHER ORGANIC MULCHES

In addition to yard waste, there are a variety of other organic mulches from which to choose. The following are some of the most common organic mulches, often available at garden centers and other outlets. Note that weed seeds can be introduced into your garden with hay, straw, or strawy manure mulches. If you use corncobs or grainy hulls, make sure they are free of grain or seeds.

Bark (hardwood)
Shredded hardwood bark, a byproduct of the paper and lumber industries, is one of the most popular and best mulches. Its pH is slightly alkaline, but you can manage this problem by adding 3 pounds of elemental sulfur for every cubic yard of bulk bark or for every 100 square feet of bed area.

Bark (softwood)
Chunk pine, fir, and redwood barks are the most popular types. This material is acidic and does not require additives to change its pH. Softwood bark is more resistant to decay than hardwood bark. It is available in a variety of sizes that fit many landscape needs.

Buckwheat hulls
Buckwheat hulls are fine-textured and may blow around if used in windy areas. This mulch is long-lived and has a neutral color that makes it satisfactory in landscape plantings. Occasionally there is a slight odor problem during hot, humid weather.

Cocoa-bean hulls
This byproduct of the chocolate industry can often be found in garden centers. Cocoa-bean hulls have good color for use on the landscape. Stir this mulch occasionally because it tends to pack down.

Corncobs (crushed)
Corncobs can be colored and used for special purposes in the landscape. Many consumers prefer the weathered, dark appearance of aged cobs to the light color of fresh ones.

Hops (spent)
These may be available from local breweries. They have excellent color and are nonflammable. The odor of fresh material may be offensive, but it subsides in a few weeks. You can get rid of the odor by using aged or composted hops.

Manure (strawy)

Strawy manure makes an excellent mulch for use in gardens if partially decomposed. To reduce the heat created during decomposition, aerate manure before using it.

Peanut hulls

This is an excellent, attractive mulch that can be obtained in garden centers near peanut-processing areas.

Peat moss

Peat moss is one of the most commonly used mulches. It has a classy look when used properly, but its cost is often prohibitive when you need to cover large areas. Although various particle sizes are available, the coarse grade is recommended for use as a mulch.

Pecan shells

This is a good, long-lasting mulch with pleasing color and texture.

Pine needles

Pine needles make a light, airy, attractive mulch. They are recommended for use around acid-loving plants. However, leave pine needles beneath pine trees; don't remove them.

Sawdust

This is a common mulch in areas where it is available. But use it as a mulch of last resort, because coarser mulches are preferred. Also, keep in mind that its decomposition will cause a nitrogen deficiency unless you apply fertilizer regularly. Aged sawdust is preferable to fresh sawdust.

Straw

Straw is used for winter protection and as a summer mulch in vegetable gardens. It is highly flammable and should not be used in high-traffic areas.

Prepared with David Williams, horticulturist, UI Cooperative Extension Service.

33 Dispose of tree residues

The unkindest cut

If you have had the bad fortune to lose a large tree on your property, you already know that once the trunk has been felled—by humans, machines, ice, wind, or lightning—you're left with the problem of what to do with all the leaves, branches, and other wood.

Some, of course, can be allowed to dry and then burned in fireplaces, wood-burning stoves, and outdoor barbecue pits, if local ordinances permit. However, unless you can hire a local firm with a tree grinder or stump chipper, the remaining green branches and twigs pose a challenge because more and more cities no longer collect "green" or landscape waste.

Your first avenue for assistance should be city hall itself. Call your local public works department and ask if there is a yard-waste reclamation facility in your area. You may have to borrow or rent a truck and pay a tipping fee to get rid of those twigs and tree limbs, but some municipalities offer yard-waste recycling services at no individual charge for taxpayers.

Other possible sources of information on dealing with debris from a felled tree are local landscaping services and firewood suppliers. These businesses may want your wood and be willing to haul it away for free or for a minimum charge.

If all else fails—perhaps you live in the country or discover such services are lacking in your locale—you may want to consider renting a chipper/shredder and doing the job yourself. Chipped material should be stockpiled and allowed to weather for at least two months before being used as mulch material. This aged material often is referred to as "composted wood chips."

If you happen to be establishing a natural habitat area in your yard, you also might want to consider placing segments of larger logs and limbs in that area. They can serve as "nursing logs," allowing young seedlings to grow in the soft organic matter that was once tree trunk.

Reclamation programs

You may want to encourage city officials to establish a yard-waste reclamation program of some kind, at least for special needs such as Christmas tree disposal. Many municipalities now offer tree pickup, chipping, and redistribution of wood chips for homeowner use. Some programs are self-supporting, with fees charged for wood chips covering the pickup and grinding costs.

What about leaves?

Two good uses for leaf waste—composting and mulch—are covered in Chapters 31 and 32, respectively. But that still leaves the question of leaf burning, an issue that has generated its share of heat in town meetings.

Leaf burning is banned in many communities because of the misery it can create for certain people, particularly those with asthma and bronchitis. One of the contributing causes to asthma is poor ambient air quality. So the bottom line is: avoid leaf burning, even if your community allows it.

If you still want to burn leaves, follow these guidelines:

- Check with your local law enforcement agency or fire department to find out whether any restrictions exist. Some communities may allow leaf burning but only during certain times and under certain weather conditions.
- Even if you do not suffer from asthma or bronchitis, consider wearing a surgical mask during burning to filter out floating particles.
- Do not burn leaves of poison ivy, poison oak, or poison sumac. Burning leaves of these plants can cause allergic reactions.

Prepared with Patrick Weicherding, urban forestry specialist, UI Cooperative Extension Service.

Protecting your drinking water

Water usually doesn't require much effort on your part. You turn the faucet, and there it is. What's more, if your water comes from a city supply, a lot of the work to keep it safe is done for you. Just keep in mind that a water company can't do everything. It can't prevent the leaching of chemicals from your pipes, and it can't tailor water to meet all tastes.

The following section makes clear what things you can or need to do yourself to ensure your water's quality. There are ideas here for those who receive their water from a municipal supply and those who receive it from a private well.

34 Know when and how to test your water

Should you have your water tested?

If you have a private water supply, you are your own regulatory agency. You are responsible for the quality of water that your family and guests drink.

However, contaminated water does not always look, taste, or smell different than safe drinking water. That's why you need to test private water supplies at least once a year—more often if problems arise.

People who get their water from a public or municipal supply have more protection because these supplies are governed by federal and state standards and are tested on a routine schedule based on the population size. But this does not mean people who use public water supplies never have reason to test. It is possible that corrosive water (water that erodes metal fixtures) or other factors can cause pipes in your home to leach contaminants, especially metals, into your water supply.

The following guidelines describe conditions in which you should consider testing your water. The first section applies to people with a private or public supply. The second section generally applies only to someone with a private water supply.

Private or public supply

Reasons to test:

- Water has an objectionable taste or smell.
- Your household plumbing contains or may contain lead pipes, lead solder joints, or brass fittings and fixtures (which have the potential to leach lead).
- The water leaves scaly residues and soap scum, or it decreases the action of soaps and detergents.
- The pipes or fixtures show signs of corrosion—a wearing away of surface metals.
- You are considering the purchase of water-treatment equipment, such as a water softener, iron-removal system, or water filters.
- You want to check the performance of water-treatment equipment that already is in use.
- You have recurrent incidents of gastrointestinal illness (diarrhea, vomiting, stomach cramps, nausea) that cannot be explained.

Private supply only

Reasons to test:

- You are buying a home and wish to evaluate the safety and quality of the water supply.
- The water stains plumbing fixtures and laundry.
- The water appears cloudy, frothy, or colored.
- Water-supply equipment (pump, chlorinators, water heater) wears rapidly.
- You are pregnant, are anticipating a pregnancy, or have an infant less than 6 months old.
- You have a new well and want to evaluate it.
- Your well does not meet construction codes.
- Your well is within or close to a livestock confinement area. Check with your state environmental agency to determine proper well setback distances from livestock confinement areas.
- Your well has a submersible brass pump in it—either new or old. Brass pumps have the potential to leach lead.
- You have mixed or loaded pesticides near the well.
- You have spilled pesticides or fuel near the well.
- You have had backsiphoning problems—incidents in which pesticide backs up from a sprayer through a hose and into a well.
- Your well is located near a known operational or abandoned gas station or fuel storage tank (buried or above ground). Testing is particularly crucial if the tank has been known to leak, or if the water smells or tastes like fuel.
- Your well is close to any of the following: retail chemical facility, gravel pit, coal mine or other mining operation, oil or gas drilling operation, dump, landfill, junkyard, factory, dry-cleaning operation, road-salt storage site, or heavily salted roadway.
- Your well is near a septic tank or septic system's absorption field. Check with your state to determine proper well setback distances from septic systems.
- You have a sand-point well, or a large-diameter dug or bored well. (These wells are more vulnerable to contamination than other wells.)
- Your well is shallow (less than 50 feet deep) and one of these conditions exist: (a) the soil is sandy or (b) bedrock or sand and gravel are fewer than 10 feet from the surface.

If any of these conditions exist, consult your state or local public health department, or a private testing lab to determine which tests should be performed on your water.

Who should do the testing?

Your local health department may do simple baseline testing for bacteria or nitrate. For more extensive testing and results, especially for pesticides and other inorganic substances, consult your state public health department, state water regulatory agency, or a laboratory certified by your state or by the U.S. Environmental Protection Agency.

How to read water test results

Water test results express the concentration of different chemicals in different ways, depending on whether they are minerals, pesticides, or other compounds.

Minerals. Water tests express the concentration of most minerals in either "parts per million" (ppm) or "milligrams per liter" (mg/L). Don't let this confuse you; one part per million is equal to 1 milligram per liter.

Pesticides. Pesticides are rarely found in water at concentrations as high as parts per million or milligrams per liter. Therefore, they are usually reported in "parts per billion" (ppb) or "micrograms per liter" (µg/L). A microgram is equal to $1/1{,}000$ of a milligram, and a part per billion is $1/1{,}000$ of a part per million.

If it is difficult to understand what is meant by these extremely small measurements, here are some examples that may be easier to picture:

- One part per million is the equivalent of a teaspoon of a contaminant dissolved in a pool of water 2 feet deep, 10 feet wide, and 12 feet long—about the area of a small- to medium-sized bedroom filled 2 feet high with water.
- One part per billion is the equivalent of a teaspoon of a contaminant dissolved in water that could cover an acre of land to a depth of 5 feet. (An acre is roughly 209 feet by 209 feet.)

Although these images emphasize how small the amounts being measured are, keep in mind that long-term exposure to small amounts may be of concern. One part per billion of a pesticide residue in 1 liter of water sounds small, but it represents billions of pesticide molecules.

Other compounds. For compounds other than minerals or pesticides, the results of a water analysis may be expressed in different forms of measurement. For instance, water hardness may be expressed in "grains per gallon," whereas the corrosion index simply estimates whether water is corrosive or not corrosive.

For more information on testing water for nitrates, bacteria, or pesticides, see Chapters 35 and 36.

Prepared with Michael Hirschi, soil and water specialist, and Mel Bromberg, drinking water and health specialist, UI Cooperative Extension Service.

HOW TO COLLECT A WATER SAMPLE

If a state or local agency is testing your water, a representative is often willing to come to your home to collect the water sample. But if you are sending a water sample to a private lab, you may need to collect the sample yourself.

Fortunately, the job is simple and straightforward. Below are a few general guidelines on collecting a water sample, as well as additional guidelines when testing for bacteria. The procedure can vary, depending on the contaminant being tested or the lab being used, so follow the lab's directions. Labs should tell you what kind of container to use in collecting a sample, and most will provide you with an appropriate container.

Tips for collecting a water sample:
- Let water flow for five minutes before sampling.
- After filling the sample bottle, mark the date and time of collection with an indelible marker on the outside of the bottle and on any lab paperwork. Then, find out from the post office whether sending the sample via priority mail will get it to the lab within forty-eight hours.
- If you live near the testing facility, consider hand-delivering the sample.

Additional tips for collecting a water sample to test for bacteria:
- For the most accurate results, samples should arrive at the laboratory within thirty hours. So, if possible, collect water samples as near as possible to post office "mail-out times"—perhaps early in the work week (Monday, Tuesday, or Wednesday). If a sample is more than forty-eight hours old, it should not be analyzed.
- Take the sample from a nonthreaded fixture, such as a bathtub spout.
- Do not take the sample from a fire or yard hydrant. Also, avoid faucets leaking at the handle and faucets where food or beverages are dispensed or prepared. If a faucet has attachments, such as a hose or aerator, either avoid the faucet or remove the attachments before collecting the sample.
- Wash your hands thoroughly with hot water and soap before collecting the sample.
- Do not touch the inside of the lid or lay the cap down while collecting the sample.
- Leave ¾ inch to 1 inch of air space at the top of the bottle.

Prepared with Michael Hirschi, soil and water specialist, and Mel Bromberg, drinking water and health specialist, UI Cooperative Extension Service.

35 Test your water for coliform bacteria and nitrate

A tale of two contaminants

Two of the most common water contaminants are coliform bacteria and nitrate. Municipal water systems regularly test for these contaminants, but if you have a private well, testing for them is up to you. The good news is that many state and county departments of public health will test for bacteria and nitrate for free or for a nominal fee. Check with your local or state health department to determine whether the service is available and to find out about the proper procedures.

Annual testing for coliform bacteria and nitrate is a good idea, especially after a heavy spring or summer rainstorm and whenever your well is flooded. You should also disinfect the well and test the water any time you open the well—for example, whenever you repair or replace an old well or pipes, and after installing a new well or pump. Opening the well can introduce bacteria into the system.

If you are planning to have a baby, already expecting a baby, or have a baby less than 6 months old, test the well water for nitrate as soon as possible. Excessive nitrate levels in the water can be harmful, sometimes fatal, to infants less than 6 months old. (See the sidebar, "Blue-baby syndrome.")

Bacterial contamination

Common sources of bacteria are livestock waste, septic systems, and surface water that gets into the well. Testing for coliform bacteria is important because it is an "indicator organism," which means that its presence may indicate the existence of other harmful bacteria in your water supply. Using an indicator organism is necessary because testing for all harmful bacteria would be difficult and expensive.

If your water test shows the presence of coliform bacteria, your water has some degree of contamination. Although the state cannot require someone with a private water supply to find a remedy, you are urged to treat the problem for your own health protection and that of your family and guests. The health department will recommend that you disinfect your water-supply system and then submit another sample for analysis.

The local department of public health, a licensed well driller, or a pump repairperson can further explain the correct techniques for disinfecting a well. In the meantime, use bottled water until you are able to bring bacteria levels under control.

Occasionally, public water supplies become contaminated with bacteria, and the supplier issues "boil orders." Boiling water is an effective way to kill pathogens (disease-causing organisms). First, you need to get the water up to a "rolling boil"—which can take up to three minutes. Once water reaches a rolling boil, continue to boil for another minute.

Nitrate contamination

Common sources of nitrate in groundwater are fertilizers, septic systems, livestock waste, and naturally occurring nitrate in the soil. A water-testing lab will describe nitrate concentrations in one of two ways. The lab may describe the nitrate concentration as the amount of actual "nitrate" or as the amount of "nitrate-nitrogen." A nitrate concentration of about 44 parts per million (ppm) is the equivalent of a nitrate-nitrogen concentration of 10 ppm.

If unacceptable nitrate levels are found in your water, do not boil the water. *Boiling water does not eliminate nitrate.* In fact, it causes some of the water to evaporate, which *increases* the concentration of nitrate in the remaining water. Use bottled water until you can treat the well water, eliminate the pollution source, or make repairs (if there is a problem with well construction).

Here are further guidelines to using water that contains nitrate:

Nitrate guidelines

Nitrate (NO_3)	Nitrate-nitrogen (NO_3-N)	Interpretation
parts per million		
0 to 44 ppm*	0 to 10 ppm*	Drinking water standard level. Safe for humans and livestock.
45 to 176 ppm	11 to 40 ppm	Water should *not* be consumed by infants under 6 months of age and pregnant women. Generally safe for other adults—as long as nitrates are the only problem. (In some cases, high nitrates can indicate other problems, such as bacterial contamination.)
More than 176 ppm	More than 40 ppm	Water should not be consumed at all.

*This is the maximum contaminant level (MCL)—the federal standard that public water supplies cannot exceed.
SOURCE: *Nitrates in Groundwater,* Kansas State University Cooperative Extension Service.

Prepared with Michael Hirschi, soil and water specialist, and Mel Bromberg, drinking water and health specialist, UI Cooperative Extension Service.

BLUE-BABY SYNDROME

The causes
Few cases of blue-baby syndrome (more properly known as nitrite poisoning) have been reported recently in the United States, but the problem is potentially fatal, so it should not be overlooked. Many doctors believe that nitrite poisoning is much more widespread than statistics indicate, for they say that blue-baby syndrome is often mistaken for other illnesses.

Methemoglobinemia (the technical name for nitrite poisoning) is usually the result of babies being fed formula made with well water containing high levels of nitrate. Have your wells tested for nitrate immediately if you have infants less than 6 months old or are planning to have a baby.

At birth, an infant's stomach is not developed enough to effectively prevent nitrate from being converted into nitrite. Nitrite enters the bloodstream and changes hemoglobin—an oxygen carrier—into methemoglobin, which cannot transport oxygen. As a result, the baby's body has a more difficult time transporting oxygen through its blood. In severe cases, the syndrome can cause death.

Infants may be more sensitive to nitrite poisoning if they are predisposed to diarrheal conditions or gastrointestinal upsets.

Physicians recommend that infant formula be mixed with tap water only if the water contains less than 44 parts per million of nitrate, which is the same as 10 parts per million of "nitrate-nitrogen." Some doctors also recommend that pregnant women avoid water high in nitrate because it may affect the unborn baby.

Symptoms
- Unusual blue skin color, similar to the color of blood vessels located close to the skin. The blue skin color may be especially noticeable around the eyes, mouth, and fingernails. Important: if a baby's skin turns blue, it is necessary to rush the child to the hospital because that means oxygen levels in the bloodstream are decreasing rapidly.
- Vomiting or diarrhea.
- Chocolate-colored blood.

Treatment
Be sure to consult a physician if you suspect nitrite poisoning of your baby; let the physician make the diagnosis. If the condition is diagnosed

early enough, you can simply use bottled water in the infant's formula to stop the syndrome.

If the physician diagnoses the problem in later stages, treatment may be required in addition to using bottled water to prepare the infant's formula. If diagnosed in time, the problem is easily corrected by proper medical procedures.

Don't boil water

It's important to know that boiling water does not eliminate nitrate. In fact, boiling water causes some of the water to evaporate, which decreases the volume of water and *increases* the concentration of nitrate in the remaining water.

To remove nitrate from water, use known treatment techniques, such as distillation or reverse osmosis. However, these systems can be expensive, so your best option may be to use bottled water until you solve the nitrate problem and the testing agency (or your doctor) advises you that the tap water is safe for your family. The safety of the water depends on the age of your children and the level of nitrate detected.

Prepared with Michael Hirschi, soil and water specialist, and Mel Bromberg, drinking water and health specialist, UI Cooperative Extension Service.

36 Test your water for pesticides—if necessary

Should you test for pesticides?

If your drinking water comes from a community supply—either a public supplier or a private company—you have little reason to test for pesticide contamination. Community water supplies are required to test for eighty-three contaminants (including many common pesticides) on a regular basis and must inform customers when a detection exceeds health standards.

However, if your water comes from a private well, you may have good reason to test for pesticides, especially in these cases:

- Someone in the immediate area has confirmed pesticide contamination in a private well.
- A commercial pesticide distributor is located nearby.
- You have a shallow, large-diameter well, which is more easily contaminated than deep, small-diameter wells.
- You are aware of pesticide mixing, spills, or tanks being emptied within a few hundred feet of your well.

Before you test, find out what types and brands of chemicals are applied on the surrounding land or mixed at nearby commercial facilities. Testing for pesticides is more expensive than testing for nitrate and bacteria, so you want to have a good idea of what you're looking for. Then find a certified laboratory that will run those tests.

If your well is particularly vulnerable to contamination, you should test for pesticides more than once. You may need to test periodically throughout the year to get an accurate picture of the problem. Because testing for pesticides can be expensive, carefully consider whether it is needed, and always ask for a cost estimate first.

Finally, consider having your water tested for "degradation products" as well—assuming you can find a lab that is set up to make these tests. Degradation products are chemicals that form when a pesticide breaks down, a natural process that can be caused by microorganisms in the soil. Degradation products usually are less toxic than the original pesticide, but sometimes they remain as toxic. Unfortunately, not all labs are equipped to test for them.

Some labs will test for specific degradation products as standard practice when testing for certain pesticides, so it is a good idea to ask.

Who can do the testing?

State environmental agencies or the local and regional offices of the department of public health will have information on where to have water tested for pesticides and may even be equipped to do the testing. Other testing facilities can be found in local engineering firms, water-treatment companies, and laboratories at local universities, especially in the departments of chemistry, agronomy, toxicology, or natural resources. Use a laboratory certified to test for the chemical compounds you're concerned about.

For more information on how to locate a public or private lab in your area, check the Yellow Pages or contact your local Cooperative Extension Service office, public health department, Natural Resources Conservation Service, or state environmental protection agency.

It is usually best to have the water tested by a public or private laboratory that does not sell water-treatment devices. Also, beware of door-to-door water-treatment salespeople who perform on-the-spot water tests. These tests are not very accurate. Once you have found a certified testing laboratory, find out how much it charges. Sometimes it is cheapest to have the lab "screen" one water sample for all of the chemicals you are looking for. Screening is an economical way to look for many chemicals, but it doesn't provide detailed information. Screening can detect contaminants or classes of compounds, but it cannot accurately determine their concentrations.

What if contamination is found?

If the test is positive for pesticides, you may want to immediately retest your well. Changes in rainfall, pesticide use, or water withdrawal can cause wide variations in the levels of pesticides found in your well. A second test may give you a better overall idea of what is in your water. Once any amount of pesticide is found in your water, seek advice from your local or state health department on actions to take and whether you need to do follow-up testing.

If pesticides consistently show up in your drinking water, it is important to contact the state or local public health department to find out the short- and long-term health effects. Another source of help is the National Pesticide Telecommunications Network, which has a toll-free number: (800)858-7378. The network system operates twenty-four hours a day, every day of the year. Generally, though, it's better to get help from authorities closer to home rather than at a distance.

In addition, if you confirm a problem, don't forget to contact neighbors who may be affected by the chemical contamination.

Health advisories

Once you know which chemical is in your water and how much was detected, find out if a health advisory summary is available from the public health department or the U.S. Environmental Protection Agency (EPA). Health advisory summaries tell the health effects of different chemicals in water. They also give the "lifetime health advisory level"—the concentration level thought to represent a reason for concern. When the level of a pesticide in drinking water is at or below the lifetime health advisory level, you can consume the water every day for an entire lifetime with a very low probability of increasing your health risks.

Health advisory levels are *not* legally enforceable standards. They serve as guidelines, and they contain a margin of safety. In general, each advisory level is set 1,000 times lower than the level known to cause a health effect. Unfortunately, health advisory levels have not yet been established for all chemicals. If a health advisory is not available, talk to your state health department or the U.S. EPA for health guidelines.

The two-page health advisory summaries used by the U.S. EPA describe both the noncancer health risks and the cancer risk from pesticides and other chemicals. If the chemical is classed as "not a carcinogen" or a "possible carcinogen," the health advisory summary will state the lifetime health advisory level. If the chemical is classed as a "probable carcinogen," the advisory will explain that risk.

What else do advisories provide?

Health advisory summaries also explain how a chemical is used, what specific health effects can occur, and what actions you should take if the chemical shows up in your water supply.

The likelihood that you will experience health effects depends a lot on your health, how long you have been and will be exposed to the water, how much the chemical is above or below the health advisory level, and whether any additional contaminants are present. Exposure to several hazardous chemicals can add to or sometimes multiply the effect of any single chemical.

How do you obtain health advisories?

You can obtain health advisory summaries from the U.S. EPA by contacting the Safe Drinking Water Hotline, Monday through Friday, 8 a.m. to 4:30 p.m. Central Time. The number is (800)426-4791. The hotline also provides details on drinking-water quality and water-treatment methods. Other sources for health advisory summaries are your state health and environmental agencies.

In addition, concerned individuals should consult with their department of public health and perhaps their physician to interpret results and decide what to do.

> **✓ CHECK IT OUT**
> For information on water-treatment methods: page 188.

Prepared with Michael Hirschi, soil and water specialist, and Mel Bromberg, drinking water and health specialist, UI Cooperative Extension Service.

Des Moines

Much of the Des Moines Water Works was under water during the 1993 floods.

The ins and outs of drinking water: What goes into it, what's taken out

If anyone can appreciate the value of tap water, Iowans from Des Moines can. During the historic Midwest floods of 1993, some of the pumps at the Des Moines Water Works ended up under 21 feet of water, cutting off drinking water for 19 days.

It was a long 19 days.

Except during such catastrophes, it's easy to take tap water for granted and give little thought to where it comes from. But what really goes on as water moves from the source to your kitchen table? The Des Moines Water Works serves as a good example of this process.

Source
Des Moines is somewhat unique in having three sources of water: two rivers (the Raccoon and Des Moines) and a groundwater source.

Pretreatment basins
Large pumps lift the water from both rivers into two pretreatment basins. There, dirt and debris are removed; then powdered activated carbon and potassium permanganate are added for taste and odor control.

Water treatment at the Des Moines Water Works

Treatment chemicals

1. **Alum or ferric chloride** Causes small particles to clump together.
2. **Activated carbon and potassium permanganate** Control taste and odor.
3. **Lime and soda ash** Soften water.
4. **Carbon dioxide** Adjusts the pH level.
5. **Polyphosphate** Keeps filters clean.
6. **Fluoride** Prevents tooth decay.
7. **Chlorine** Disinfects water.

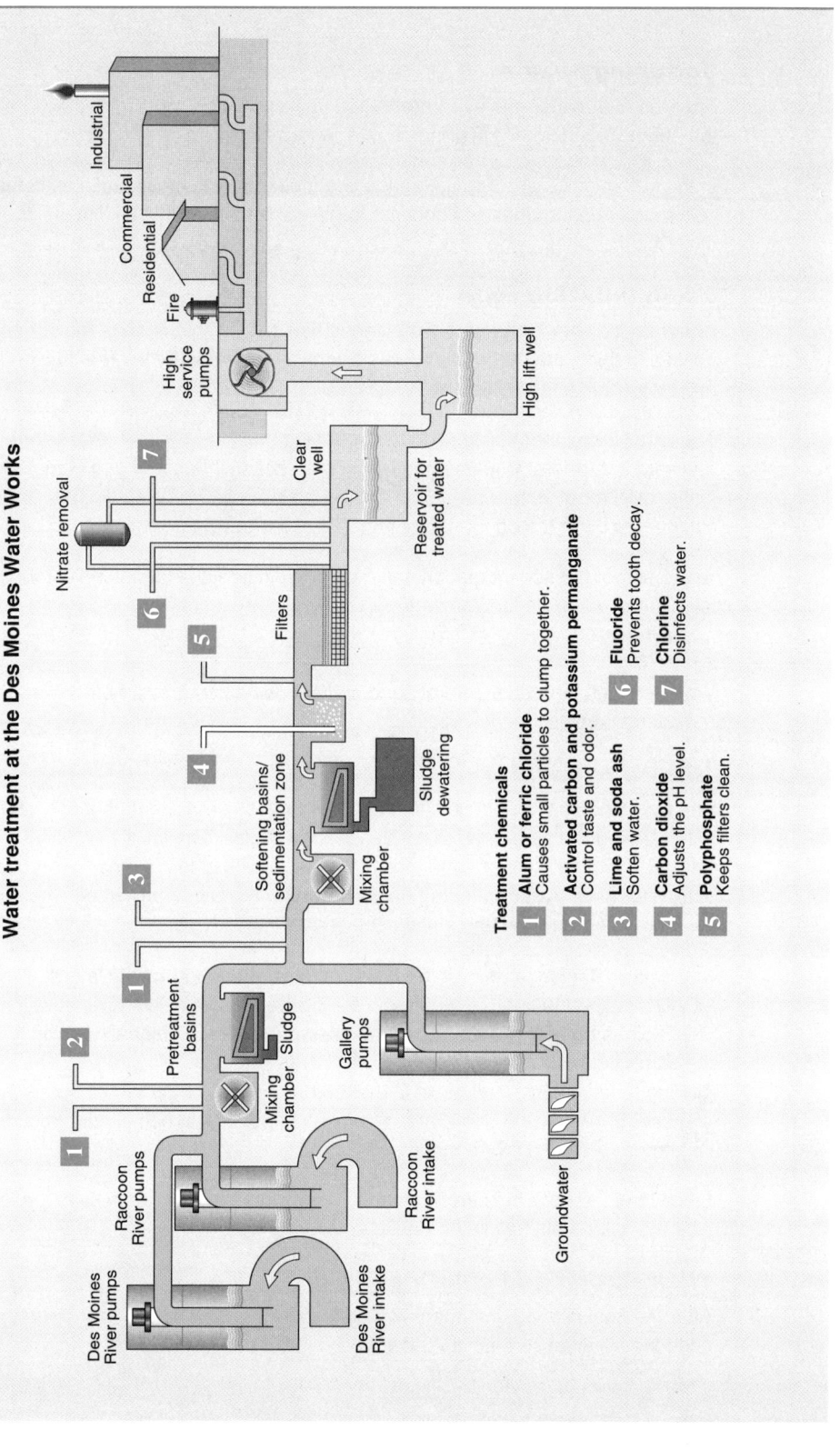

183

Softening basins

After the river water has been pretreated, it is combined with water drawn from the underground source. Then it moves to four softening basins.

In the softening basins, ferric chloride is added, causing small particles in the water to clump together and become larger particles. In addition, lime is added to soften the water.

Sedimentation zone

Water passes slowly through a sedimentation zone within each of the softening basins. In this zone, large particles—which were created by the ferric chloride and lime—fall to the bottom of the basins.

The lime sludge that builds up in the softening basins is collected, dewatered, and sold to farmers, who use it to adjust their soil pH. According to Gary Benjamin, director of water production at the Des Moines Water Works, the lime sludge has been tested for safety and "passes with flying colors."

At the end of the sedimentation zone, carbon dioxide is added to adjust the pH level of the water and to complete the softening process.

Filtering

Water passes through layers of sand and various-sized gravel to remove any remaining particles in the water.

Nitrate removal

The Des Moines Water Works is one of the largest water-treatment operations in the United States to have a nitrate removal facility. The facility removes nitrate by exchanging the nitrate ions in the water with chloride ions—a process similar to home water-softening units. However, it is used only when nitrate in the raw water from the rivers rises above allowable levels.

In the past, Benjamin says, if the Raccoon River had high nitrate levels, they simply switched to their backup source—the Des Moines River. Because the Des Moines River has a flood-control reservoir, nitrate problems don't hit it at the same time that they hit the Raccoon River. But even this strategy was beginning to fail as the nitrate levels and the duration of the nitrate episodes increased. Back in 1990, the Water Works had to send out nitrate warnings to customers on and off for a month.

Since they installed the nitrate removal facility in 1991, Benjamin says the levels have been held below the health standard.

Fluoridation and chlorination

After the final filtering, fluoride is added to prevent tooth decay, and chlorine is added to kill disease-causing bacteria.

Distribution and storage

After water is processed, it travels through a series of "feeder mains" for residential, commercial, and industrial use. The Des Moines distribution system consists of more than 800 miles of pipe set out in a gridlike pattern like an underground superhighway.

The final step is for water to come rushing out of your tap—as long as another hundred-year flood hasn't fouled the works.

37 Know the signs of contaminants in drinking water

Tan water?

A mother in West Virginia looked at the tan water running out of her sink faucet and said, "Better than usual. Much better. Sometimes it comes out so thick it won't flow down the drain—just piles up in the sink."

As this example from an article in *National Geographic* indicates, some water problems are not hard to spot at all. But even when a problem is obvious, the cause of the contamination and the solution are not so clear.

The following list may help you identify which contaminants to test for. It concentrates on those contaminants that leave sensory clues—taste, smell, and color. The list does not provide solutions, but it may help you select the proper water-treatment method (see Chapter 38). Also, be aware that if some of these chemicals are found at low levels in the drinking water, they may not leave noticeable signs. Even at low, unnoticeable levels, some of them can cause problems.

To find out what potential hazards these chemicals may pose, contact your local or state public health department.

Contaminants that leave clues

Visual clues	Possible causes
Blackening, pitting of sinks and fixtures	Hydrogen sulfide (gas), manganese
Blue-green stains on sink and porcelain fixtures	Corrosion from copper, brass, and other metals
Cloudiness	Dirt, sand, clay, organic matter (also known as turbidity), methane gas
Foamy water	Foaming agents such as detergents, dilute sewage
Milky-colored water	Methane gas, any type of particles
Reddish brown water	Iron
White deposits in pipes, water heater, tea kettle; soap and scum in the bathroom	Hard water caused by dissolved minerals, such as calcium and magnesium
Yellow water	Tannins from organic soil and vegetation

Odor clues	Possible causes
Bleachlike odor, chlorine odor	Chlorine, chloramines
Detergent smell	Foaming agents
Musty, earthy odor	Organic matter
Oil, gas smell	Gasoline or semivolatile compounds
Rotten egg smell	Hydrogen sulfide
Sweet pungent smell	Volatile organic compounds, semivolatile compounds |//
Taste clues	**Possible causes**
Gritty, abrasive	Fine sand and grit (sediment)
Metallic taste	Corrosion from iron, manganese, copper, lead, and other metals; also, sodium chloride and sulfates
Sharp chemical taste	Pesticides, volatile and semivolatile compounds

SOURCE: Adapted with permission from Water "Sense" Wheel™, © 1995, EHMI. All rights reserved. Environmental Hazards Management Institute, P.O. Box 932, Durham, NH 03824, (603)868-1496.

Prepared with Mel Bromberg, drinking water and health specialist, and Michael Hirschi, soil and water specialist, UI Cooperative Extension Service.

38 Select effective water-treatment methods

Treatment time

Taste. Odor. Hardness. Contamination. These are four of the most common reasons why people install water-treatment systems in their homes.

Those who receive their water from municipal water supplies usually turn to water-treatment systems because of taste or odor problems (often caused by chlorine) or water hardness. Those who receive their water from a private well may have similar concerns; but they also may have an added reason for installing a water-treatment system. For private well owners, contamination (especially bacterial contamination) is a greater risk than for those on municipal supplies.

If you identify contaminants in your well and determine that levels exceed standards or guidelines, first pinpoint the source and try to correct the problem. You may be able to solve the problem by taking a look at the well's construction and determining whether repairs or replacement is in order. If correcting well deficiencies isn't the answer, consider a water-treatment method.

Make sure you know exactly which contaminants you want to get rid of, as well as your financial limits. Consult with local officials and water-treatment specialists at state agencies or universities. If you are thinking about having your water tested by someone in the business of selling treatment devices, keep in mind that the company may have a conflict of interest. Also, note these points:

- No single water-treatment system corrects all water-quality problems.
- All systems have limitations and life expectancies.
- All systems require routine maintenance, monitoring, or both.

The following guide is intended only to give you a general idea of the available technology. This guide tells you which contaminants each system can remove. But remember, each system does not work with equal efficiency on all of the contaminants listed.

Activated carbon filter

What it removes. Many volatile organic chemicals, some pesticides, radon gas, hydrogen sulfide, and mercury. Also reduces odor, color, and taste problems (such as residual chlorine).

How it works. Water is filtered through carbon granules. Contaminants attach to the carbon and are trapped.

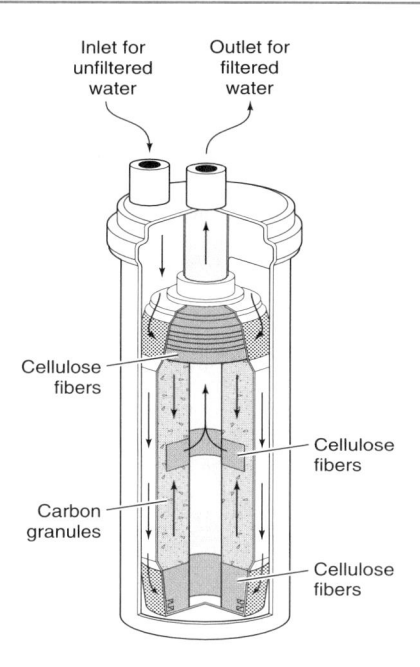

With an activated carbon filter, water enters the top of the cartridge. As water passes through carbon granules, certain contaminants attach to the surface of the material and are removed from the water. Water then moves up through cellulose filters to the outlet.

Limitations. If the filter is not replaced regularly, it will lose its ability to filter contaminants. As a result, contaminants may reenter water in amounts that are even more concentrated than before. Infrequently maintained filters can also serve as breeding grounds for bacteria.

Distillation

What it removes. Radium, odor, off-tastes, heavy metals, some pesticides, nitrate, and salt. Units with volatile gas vents can remove some volatile organic chemicals as well.

How it works. Water is evaporated, leaving impurities behind. The steam is then cooled and becomes distilled water.

Limitations. The distillation process is slow and consumes a lot of energy, making it expensive. It also consumes large amounts of water if the coolant used in the distillation process is water. Distilled water can corrode materials such as iron and copper in plumbing systems.

Air stripping

What it removes. Gases such as radon, hydrogen sulfide, and methane. Also removes some pesticides and many volatile organic chemicals, and helps treat some odor and taste problems.

How it works. Water flows down a tube while air is pumped up. Contaminants are transferred from the water to the air and vented outside.

Limitations. Air stripping is generally not practical for home use. Energy costs can increase because you repump the water after treatment and power the fan that blows air through the water. Also, units may be noisy and bacteria may grow in the water-holding tank.

Reverse osmosis

What it removes. Inorganic minerals such as radium, sulfate, calcium, magnesium, potassium, nitrate, fluoride, boron, and phosphorus. Also helps to remove salts, certain detergents, volatile organic contaminants, some pesticides, and taste- and odor-producing chemicals.

How it works. Water is filtered through a membrane. The membrane passages are smaller than the contaminant molecules.

Limitations. Under-the-sink installations are costly and take up a lot of space. Costly filter replacement. Slow and wasteful of water. Some microorganisms may be small enough to pass through the reverse osmosis membrane and colonize the holding tank.

Cation or anion exchange (water softener)

What it removes. Barium; radium; and taste-, color-, and odor-producing chemicals. Will remove dissolved iron and manganese when they are present in low concentrations. Also, anion exchange units will remove nitrate, but cation exchange units will *not*.

How it works. As hard water passes through resin beads, magnesium and calcium ions attach to the beads. They are removed from water, replaced by sodium. This softens water.

Limitations. People with hypertension or high blood pressure should consult their doctor about personal health risks associated with drinking softened water. Sodium is normally added to water during the softening process and can cause health problems for those prone to hypertension. Remember, cation and anion exchange softeners are different types of water softeners and remove different minerals.

Mechanical filtration

What it removes. Dirt, sediment, loose scale, and insoluble iron and manganese (flakes that have not dissolved).

How it works. Sand, filter paper, compressed glass wool, or other straining material clears the water of dirt, sediment, and other particles.

Limitations. Mechanical filtration does not do much to remove harmful, dissolved chemicals.

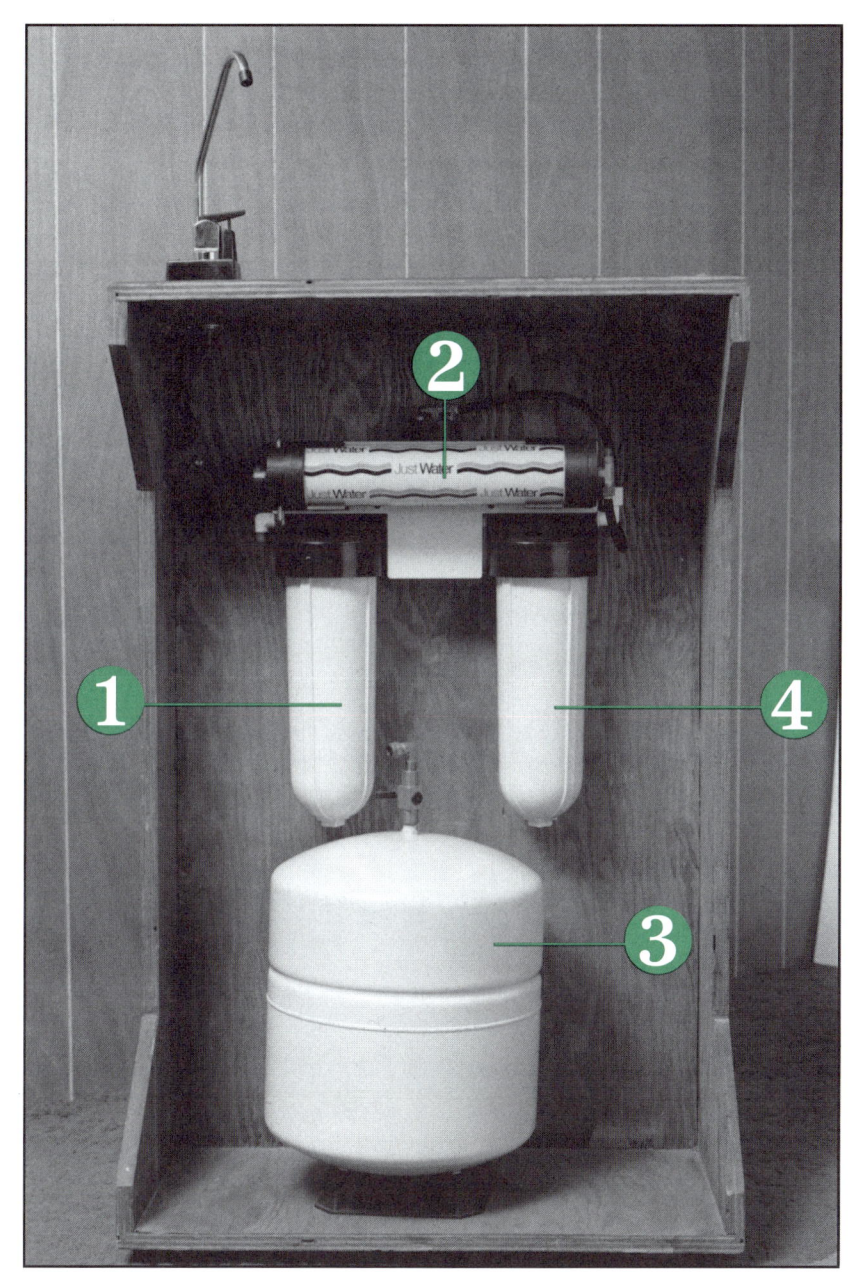

With reverse osmosis, water passes through a sediment filter ❶, which removes coarse particles that could clog the unit. Water is then forced through the reverse-osmosis membrane ❷, leaving contaminants behind. Contaminated water drains away and treated water moves to the holding tank ❸. When water is withdrawn, it moves from the holding tank, through an activated carbon filter ❹, to the faucet.

Some systems do not include an activated carbon filter, and some systems have an activated carbon filter placed **before** the reverse-osmosis membrane.

Treatment methods

Contaminants	Activated carbon filters	Air stripping	Cation or anion exchange/water softener	Chlorination	Distillation[8]	Mechanical filtration	Oxidizing filters	Ozonation	Reverse osmosis[8]	Ultraviolet radiation
Chlorine	✖									
Coliform bacteria, other microorganisms				✖				✖		✖
Color	✖		✖	✖			✖	✖		
Hydrogen sulfide	✖	✖		✖[1]			✖	✖[1]		
Inorganics, minerals, and heavy metals (lead, mercury, arsenic, cadmium, barium)	✖[2]		✖[3]		✖				✖	
Iron/manganese – dissolved			✖[4]	✖[1]			✖	✖[1]		
Iron/manganese – insoluble						✖	✖			
Nitrate			✖[5]		✖				✖	
Odor and off-taste	✖	✖	✖	✖	✖		✖	✖	✖	
Pesticides (some)[6]	✖	✖			✖			✖	✖	
Radium			✖		✖				✖	
Radon gas	✖	✖								
Salt					✖				✖	
Sand, silt, clay (turbidity)						✖				
Volatile organic chemicals	✖	✖			✖[7]				✖	
Water hardness			✖							

1. When followed by mechanical filtration or an activated carbon filter.
2. Mercury only.
3. Barium only.
4. When present in low concentrations.
5. Anion exchange units will remove nitrate. But cation exchange units will not.
6. For information on ways to treat water for specific pesticides, obtain pesticide health advisory summaries; see the section on health advisories on page 180.
7. Works for volatile organic chemicals with high boiling points.
8. While distillation and reverse osmosis work for many contaminants, they may not be the treatment of choice due to cost and maintenance requirements.

Chlorination

What it removes. Bacteria; other microbiological contaminants; and some taste-, odor-, and color-producing chemicals. Also removes hydrogen sulfide and dissolved iron and manganese when followed by mechanical filtration or an activated carbon filter.

How it works. A pump feeds chlorine into the water. The pump can dispense chlorine in direct proportion to the rate of water flow. Chlorine has a residual effect, which means it works for a while after being added to the water.

Limitations. If the system is not operated properly, it is expensive and possibly toxic. Chlorination byproducts may be harmful.

Ultraviolet radiation

What it removes. Bacteria and other microbiological contaminants.

How it works. As water passes by a special light bulb, ultraviolet radiation kills contaminants.

Limitations. May not work effectively in cloudy water or when the water flow is too fast. Unless the unit is equipped with a special meter, it is hard to know whether the system is doing the job. UV units do not have a residual effect, as chlorination does.

Ozonation

What it removes. Bacteria; other microbiological contaminants; some pesticides; and some taste-, odor-, and color-producing chemicals. Also removes hydrogen sulfide and dissolved iron and manganese when followed by mechanical filtration or an activated carbon filter.

How it works. Water is exposed to ozone gas, which destroys microorganisms.

Limitations. Equipment to generate ozone is expensive. Ozonation does not have a residual effect, as chlorination does.

Oxidizing filter (greensand filter or zeolite filter)

What it removes. Iron, manganese, and hydrogen sulfide. Also removes some taste-, odor-, and color-producing chemicals.

How it works. Contaminants are removed through filtering and chemical reactions.

Limitations. The system needs to be regenerated by pouring potassium permanganate into it. Potassium permanganate can pose a hazard to eyes and skin during handling, so always wear gloves.

Prepared with Michael Hirschi, soil and water specialist, and Mel Bromberg, drinking water and health specialist, UI Cooperative Extension Service.

WATER-TREATMENT SCAMS

What to look for
Many water-treatment scams follow two patterns. Salespeople may try to sell devices that really don't work, or they may sell devices to consumers who really don't need them. To do this, many scam artists try the following strategies:

- They take advantage of fear and ignorance about water quality instead of relying on the quality and effectiveness of their product.

- They test water in the home using their own testing kits, which are not necessarily the kind found in water labs. This leaves a consumer unsure of the accuracy of the results.

- They pretend to be taking a survey or giving away prizes in a contest. They may also pretend to be working for the government because water in the area has been "reported" as being impure.

- They claim their product has been approved or recommended by the U.S. Environmental Protection Agency (EPA). However, the U.S. EPA does not test or approve water-treatment devices. It only registers them. The National Sanitation Foundation (NSF) tests and certifies many water-treatment devices for home use. So check for the NSF seal.

- They want you to make an on-the-spot decision, instead of giving you time to shop around and think about your purchase.

What to do
Here's how to avoid becoming a victim of these types of scams:

- Have water tested by a government agency (such as the public health department) or an EPA-certified laboratory that does not sell home water-treatment devices. When you know what contaminants are in your water, carefully examine all water-treatment options.

- Rent a water-treatment system and evaluate its performance prior to purchase.

- Never give your credit card number over the phone unless you are familiar with the company.

- Read all contracts carefully.

- Call the Better Business Bureau or a consumer protection agency to find out if there are any unresolved complaints about the company.

- If you are the victim of scare tactics or misleading advertising, call or write to the Better Business Bureau, Federal Trade Commission, Cooperative Extension Service, Water Quality Association, state and U.S. EPA, local and state health departments, or local media.

- Learn about your local or private water system so you are informed when a salesperson tries to tell you about your water.

Prepared with Michael Hirschi, soil and water specialist, and Mel Bromberg, drinking water and health specialist, UI Cooperative Extension Service.

39 Prevent contamination from septic systems

The cost of malfunctions

A septic system is something you want to keep out of sight but not necessarily out of mind. If you don't maintain your septic system on a regular basis, the system could malfunction and cause:

- Contamination of groundwater and surface water
- Spread of sewage-borne diseases such as shigella, giardiasis, and, more commonly, acute gastroenteritis
- Costly damage to the home and septic system

The basic components

The most important factors in keeping a septic system operating effectively are proper soil conditions, proper sizing of the system, and homeowner maintenance. It's also important to know the basic components of a septic system. Typical septic systems have two key elements—a septic tank and an absorption system. To get an idea of how these components work, see the accompanying illustration.

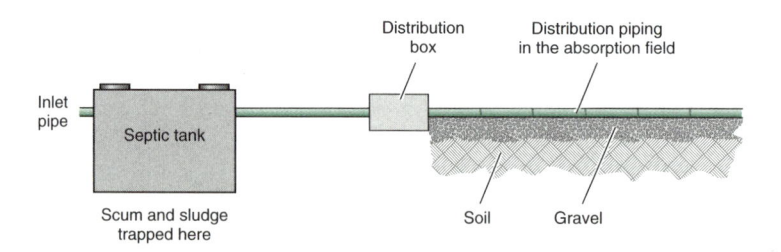

Septic tank. The septic tank is usually a concrete container that receives wastewater from your bathroom, kitchen, and laundry room. It allows heavy particles (sludge) to settle and light materials (scum) to float. In the tank, bacteria break down some waste products, allowing liquids to move into the absorption system.

The absorption system. The absorption system, or drainfield, consists of a distribution box, perforated distribution lines, and a soil area that has the capacity to accept wastewater. Wastewater from the septic tank moves through the drainfield, where harmful microorganisms, organic materials, and nutrients are removed.

Basic maintenance

To help your septic system operate correctly, follow these guidelines:
- Know the locations of all of the parts of your system, and don't run heavy vehicles over them.
- Avoid planting trees or shrubs near drain tiles. Their roots can clog drain lines.
- Divert surface runoff around the system, if possible.
- Be careful of what you dispose of in the toilet or in your drains. Household chemicals can destroy the bacteria that break down organic material in your septic tank; garbage disposals can add unnecessary solids and grease to your system; and nonbiodegradable materials can clog the absorption field.
- Conserve water whenever possible. To avoid overloading your system on any particular day, try to distribute throughout the week your laundry and other chores that require heavy water use.

> **CHECK IT OUT**
> For ideas on how to conserve water: page 283.

- Install a lint trap on the washing machine. Lint can clog the septic system.
- Monitor your septic tank annually, and have a reputable contractor pump it out every two to three years—or more frequently, if needed. Letting the tank overload with sludge reduces the time that wastewater remains in the tank. As a result, fewer solids settle in the tank, fewer solids decompose, and more solids reach the absorption field. Clogging the field with solids can result in premature failure of the absorption field and may require costly repairs or replacement.

Other systems

Systems that do not use a septic tank and absorption field are also suitable for waste treatment. Contact your local department of public health for information on installation and maintenance of other systems.

Prepared with Michael Hirschi, soil and water specialist, UI Cooperative Extension Service.

40 Know the pros and cons of bottled water

H_2O to go

In 1993, the contamination of Milwaukee's public water supply and the ensuing illnesses sent residents scrambling for alternative water sources, while Washington, D.C. made headlines the same year with its municipal water problems. In the bottled water industry, meanwhile, one supplier received widespread attention in 1990 when benzene showed up in its popular brand of water.

When you look beyond these highly publicized incidents, you are left with the basic question: "Is bottled water any better than tap water?"

Not necessarily. As an indication of how good the tap water from municipal supplies can be, about one-quarter of all bottled water originally comes from municipal water wells. However, there are two good reasons for considering bottled water:

Contaminated tap water. If contamination (biological or chemical) has been detected in your water supply at levels that exceed acceptable levels, your decision is simple. You may wish to turn to bottled water—at least until the contamination problem is solved. Public water suppliers are required to notify you whenever a contaminant exceeds acceptable levels. Private well owners should have their water tested annually for at least nitrate and bacteria; then follow appropriate guidelines if the water is contaminated.

Personal taste. If your tap water is not contaminated, the decision boils down to personal taste. For instance, some people have problems with the taste or odor of chlorine in their tap water; they may choose bottled water that is treated with tasteless, odorless ozone (check the label).

Comparing tap water and bottled water

The best way to compare your tap water with bottled water is to look at water test results. How you do this depends on your source of tap water.

If your tap water comes from a rural, private well, have the water tested by a lab approved by the U.S. Environmental Protection Agency (EPA). Your local public health department or Cooperative Extension Service office should have a listing of EPA-approved labs in your state.

Those who obtain water from private wells may have good reason to consider an alternative source, such as bottled water. Studies of private wells in many states have uncovered potential problems with a number of contaminants, including bacteria, viruses, nitrate, and pesticides.

If your tap water comes from a municipal water supply, you can evaluate its quality by requesting reports of water test results from your supplier. Public water supplies are required to monitor for certain contaminants on a regular basis. When a contamination problem occurs, suppliers must notify their customers and take whatever measures are necessary to correct the problem.

Keep in mind that trace amounts of contaminants may show up in bottled water as well. The difference is that if the level of a contaminant in bottled water exceeds federal health standards, the product cannot be sold. When a contaminant exceeds health standards in a public water supply, the water may keep flowing while the problem is being corrected.

To determine the level of contaminants in a specific brand of bottled water, request water test results from the company. Be sure to request results for the specific contaminants you are concerned about.

What can be done about chlorine taste?

If your tap water has a chlorine taste that bothers you, switching to bottled water is an option. But it is not the only one. Here are two other ways to deal with a chlorine taste problem:

Aerate your tap water. If you let your water "aerate," the chlorine taste is greatly reduced. One way to do this is to install an aerator cap on your faucet. Another idea is to put your tap water in a bottle, loosen the cap, and store it in the refrigerator. After a few hours, the chlorine gas will evaporate, and the odor and taste of chlorine will be less offensive.

Filter out the chlorine. An activated carbon filter can remove chlorine in tap water. However, bacteria can sometimes grow on these filters. To avoid bacterial growth, follow the manufacturer's directions on changing the filter.

How is bottled water regulated?

Bottled water must meet federal and state standards:

Federal standards. Bottled water is regulated by the Food and Drug Administration (FDA), which requires manufacturers to submit samples regularly for testing. Both imported and domestic bottled water must meet FDA standards, which are not much different than the U.S. EPA standards that govern public drinking-water supplies.

State standards. Individual states must enforce the federal regulations, but they can also pass stricter standards.

International Bottled Water Association. The bottled water industry regulates itself through the International Bottled Water Association

(IBWA). Bottled water producers who are members of IBWA are inspected annually by an independent laboratory. But not all water bottlers are members of the IBWA. The label may indicate whether they are.

Keep bottled water clean

If you decide to buy bottled water, make sure it does not become contaminated with bacteria after you bring it home—otherwise the added expense may go to waste.

The wet, warm, threaded cap of an unrefrigerated bottle of water is a possible place for bacteria to grow, and they will begin to grow as soon as you break the seal. If ingested, some bacteria cause gastrointestinal problems and other health risks.

To keep your bottles clean, follow these hints:

- Store the bottle in a refrigerator at temperatures above freezing but less than 50°F.
- Avoid any type of buildup in the bottle cap.
- If your bottle is refillable, make sure it is well cleaned and rinsed before refilling. If possible, recycle the old bottle and obtain a fresh, sterile, sealed bottle.

Prepared with Mel Bromberg, drinking water and health specialist, UI Cooperative Extension Service.

SPARKLING, NATURAL, MINERAL, SPRING, ARTESIAN...ARE YOU CONFUSED?

If you don't understand the difference between "sparkling natural water" and "naturally sparkling water," you are in good company. Here is a handy guide to the different *sources* of bottled water and the different *types* of water.

Sources of bottled water

Wells	Water is pumped from an underground water source, known as an "aquifer."
Artesian wells	Water flows naturally up through the well. Pumping is not necessary.
Springs	Water flows naturally to the surface. Construction of a well is not necessary.
Municipal supplies	Water comes from a privately or publicly owned water-treatment plant. Source of water can be surface water or groundwater.

Types of water

Still water	Water without "fizz." Can be either tap water or bottled water.
Sparkling water	Carbonated water with "fizz."
Sparkling natural water	Comes from a natural source. Mechanically carbonated.
Naturally sparkling water	Comes from a natural source. Naturally carbonated.
Mineral water	Contains dissolved minerals. May be sparkling or still water.
Natural mineral water	Comes from a natural source. Contains naturally occurring minerals.
Distilled or demineralized water	Has had minerals and sodium removed.

Prepared with Mel Bromberg, drinking water and health specialist, UI Cooperative Extension Service.

Protecting the indoor environment

For many people, the things that seem to haunt houses today are radon, lead, asbestos, and pesticides. And like so many frightening things, much of what makes them ominous is the unknown. Many homeowners do not know what risks these contaminants pose, how to identify problems, or how to reduce contamination.

The information that follows sheds some light on these confusing issues.

41 Prevent contamination by lead

A historical hazard

Lead has had a history of creating trouble. For instance, some people believe that lead pipes and other sources of lead contaminated the drinking water of ancient Rome, undermining the citizens' health. And in seventeenth-century Europe, many women suffered the consequences of using cosmetics that contained white lead.

You cannot see trace amounts of lead. You cannot taste it. You cannot smell it. But even minute amounts of lead can cause health problems, especially for children and pregnant women.

When lead is swallowed or inhaled, it enters the bloodstream and interferes with the formation of red blood cells. Lead can also harm the nervous system, kidneys (at very high levels), and reproductive system. In addition, it has been known to slow the development of children, damage hearing, and decrease intelligence.

Fortunately, lead exposure in the United States is decreasing, mainly because the U.S. food industry has eliminated lead solder in cans, and lead has been banned in gasoline. However, risks still exist and precautions need to be taken.

Lead in paint

The dangers from lead-based paint have been well publicized and well documented for a long time—and for good reason. The ingestion of dust from lead-based paint is still the major way that lead gets into a child's system.

Although lead-based paint continues to be used for exterior metals, it is no longer sold for household use. Nevertheless, old coats of lead-based paint may surround you on the woodwork, cabinets, and porch of your house. To find out if the paint in your home contains lead, contact the local health department for assistance. A laboratory can analyze paint chips for lead, or specialists can take readings of the lead levels in your home with an X-ray device. If you are doing renovations that will call for sanding or scraping away bottom layers of paint, it is crucial to have the older layers tested in addition to the top layer.

If you discover layers of lead-based paint, here are some options.
- Check the Yellow Pages or contact the state health department to locate licensed paint-removal companies that can deal with lead paint.
- Monitor the window sills, furniture, and carpets for paint chips and

dust. Vacuum these areas regularly after wiping them clean with a moist cloth.
- If the paint is peeling or chipping, consider covering the surface at least temporarily with wallpaper or contact paper.
- Clean floors regularly with a wet mop followed by vacuuming.
- Wash children's hands before they eat or go to bed. If they have lead dust on their hands and then stick their fingers in their mouths, they can ingest the lead.
- Prevent children from chewing on anything covered with lead paint or dust—toys, window sills, and cribs.

Lead in drinking water

Lead enters drinking water as a result of corrosion, or wearing away, of lead pipes, which were once used in some parts of the country to connect a home to the water supply. Lead can also leach from lead-based solder used to join copper pipe.

Since 1986, the amount of lead in piping has been reduced to no more than 8 percent. At the same time, lead levels in solder and flux were reduced to no more than 0.2 percent for use in plumbing that supplies drinking water.

Lead in drinking water is most often a problem in homes that are very old. Consider having your drinking water tested for lead in these situations:

- If you know or suspect that your home's water system has lead pipes or is served by a lead service connection. Homes built in the 1950s or before may have lead pipes or lead connections to the service main.
- If you have a child with excessive lead levels in the blood.

If you find yourself in either of the above situations, here are some suggested steps to follow:

1. To find out where to have your water tested, check with the state public health department office, which should have a list of certified laboratories in the area. The maximum allowable level of lead in public drinking-water supplies is 0.015 milligrams per liter, or 15 parts per billion. However, no amount of lead in water is good, and some specialists recommend that you consider taking action to solve the problem when the lead level reaches as little as 5 parts per billion.

2. If test results indicate lead contamination of your water supply, determine the source of contamination. Have a licensed plumber inspect pipes and pipe fittings for lead.

3. If lead connections or lead solders are found, one option is total replacement of plumbing lines, but that can be expensive and is usually

unnecessary. The following are some other things you can do to lessen the amount of lead in your water:

- Use only cold water for cooking and drinking. Hot water causes more lead to leach from pipes and, therefore, can contain more lead than cold water.

- Flush standing water from the system by running the cold water from two to five minutes, *until it is as cold as it will get,* before using the water at the beginning of the day. If the house has a lead service line to the water main, run the cold water for a longer time. Do this at whatever faucet is being used to provide water for drinking and cooking.

  To find out if this strategy is effective in reducing lead to acceptable levels, take water samples before and after running cold water for two to five minutes. Then have the two samples tested for lead, and compare results.

- Consider purchasing an in-home water-treatment device. A water distiller can remove lead and other contaminants, but will leave water tasting flat. Reverse osmosis units are also effective at removing lead, but they can use several gallons of water to produce one gallon of "treated" water.

  Before you buy a distiller or reverse osmosis unit, carefully investigate the product, and try to rent a unit for a month or two to see how well it works and how it affects your water bill.

Lead in food

Lead can leach into food through the use of ceramic plates, bowls, and pitchers that haven't been glazed properly. Of particular concern, according to the Food and Drug Administration (FDA), is the frequent or daily use of ceramic mugs for hot beverages. Hot beverages are more apt to cause lead to leach than cold ones. However, *acidic* cold beverages (such as fruit and vegetable juices and iced tea) also have a tendency to cause leaching of lead.

To reduce the risk of lead exposure in food and beverages, observe these precautions:

- If you are pregnant, avoid the use of lead crystalware and the daily use of ceramic mugs for hot or acidic beverages.
- Do not store acidic foods and beverages in ceramic containers.
- Do not store beverages in lead-crystal containers.
- Use antique or collectible housewares just for special display.
- Don't use items that show a dusting of chalky gray residue on the

glaze after they are washed. This could indicate that the glaze is damaged and that lead could leach into food.
- Follow label directions if ornamental plates or containers are marked, "Not for Food Use—Plate May Poison Food. For Decorative Purposes Only."
- If wine is sealed with a foil capsule, wipe the rim of the bottle with a cloth that is dampened with water or lemon juice before removing the cork. Foil wrappers on wine bottles can contain lead.

Some companies offer kits that allow consumers to test ceramicware for the leaching of lead. According to the FDA, these kits are effective in detecting ceramicware that releases *large* levels of lead, but not those that release low levels.

Testing blood-lead levels in children

Lead poisoning does not always produce symptoms. Therefore, to find out if your children have elevated levels of lead, consult with your doctor and have a blood test performed. Some experts recommend that tests be conducted when children are 6 months old and then again at 24 months. A screening test is *required* in certain states before a child enters preschool, day care, or kindergarten.

If your child has an excessive blood-lead level, you probably will want to have yourself tested as well. Also, women planning to have a baby should be tested for lead, because lead can affect the baby's size and the term of delivery.

According to the Centers for Disease Control, a person's blood-lead level should be less than 10 micrograms per deciliter. If the blood-lead level is higher, don't panic, but talk to your doctor about precautions that you can take to prevent the level from rising any higher.

Foods high in iron and calcium—such as meat, eggs, raisins, milk, cheese, fruit, and potatoes—help *prevent* the absorption of lead in a child's system. Foods high in fat and oil—such as potato chips, french fries, and other fried food—*promote* the absorption of lead in a child's system, so limit them. Finally, have children wash their hands before they eat or go to sleep so lead dust does not transfer from their hands to their mouth.

Prepared with Dawn Hentges, food-safety specialist, Mel Bromberg, drinking water and health specialist, and Michael Hirschi, soil and water specialist, UI Cooperative Extension Service.

42 Test your home for radon

Sounding the alarm

In 1984, a Pennsylvania man set off radiation alarms at the power plant where he worked. What baffled him most was that he set off alarms while *entering* the plant. As it turned out, the radiation problem was traced to his home, where radon gas had reached extremely high levels.

This incident also helped to set off alarms nationwide, increasing concern and awareness about radon. Radon had been known to be a problem for miners, but the Pennsylvania case brought to light that radon can be a health hazard in residences. According to the U.S. Environmental Protection Agency (EPA), radon gas may be responsible for anywhere from 7,000 to 30,000 lung cancer deaths per year, making it a leading cause of lung cancer in the United States—second only to cigarette smoking.

Some scientists dispute this figure, saying that it exaggerates the risk. At the core of the debate is whether studies of high-level exposures among uranium miners can be used to determine the risk posed by low levels of radon in homes. What's more, research remains divided. For instance, a recent issue of the *Journal of the National Cancer Institute* found no correlation between radon and lung cancer in *nonsmoking* Missourians. However, a study of Swedish residences, reported in the January 20, 1994, issue of the *New England Journal of Medicine,* supported the use of uranium miner studies to estimate the risk in homes.

What is radon?

Radon is created naturally in the earth—a tasteless, colorless, and odorless gas that forms when uranium in the Earth's soil decays. As radon decays, forming what are called "radon progeny," alpha particles are emitted and can damage cells in the human body.

Outdoors, radon gas is diluted in the air and poses little risk. But indoors, it can become concentrated and can accumulate to hazardous levels. Radon seeps into homes through openings such as:

- Sump pumps
- Gaps between floor-to-wall joints
- Cracks in basement walls, floors, and foundations
- Crawl spaces
- Pores in concrete block
- Cracks or pores in hollow-brick walls

Pipeline from well

Radon sources

A Crawl space

B Gaps between floor-to-wall joints

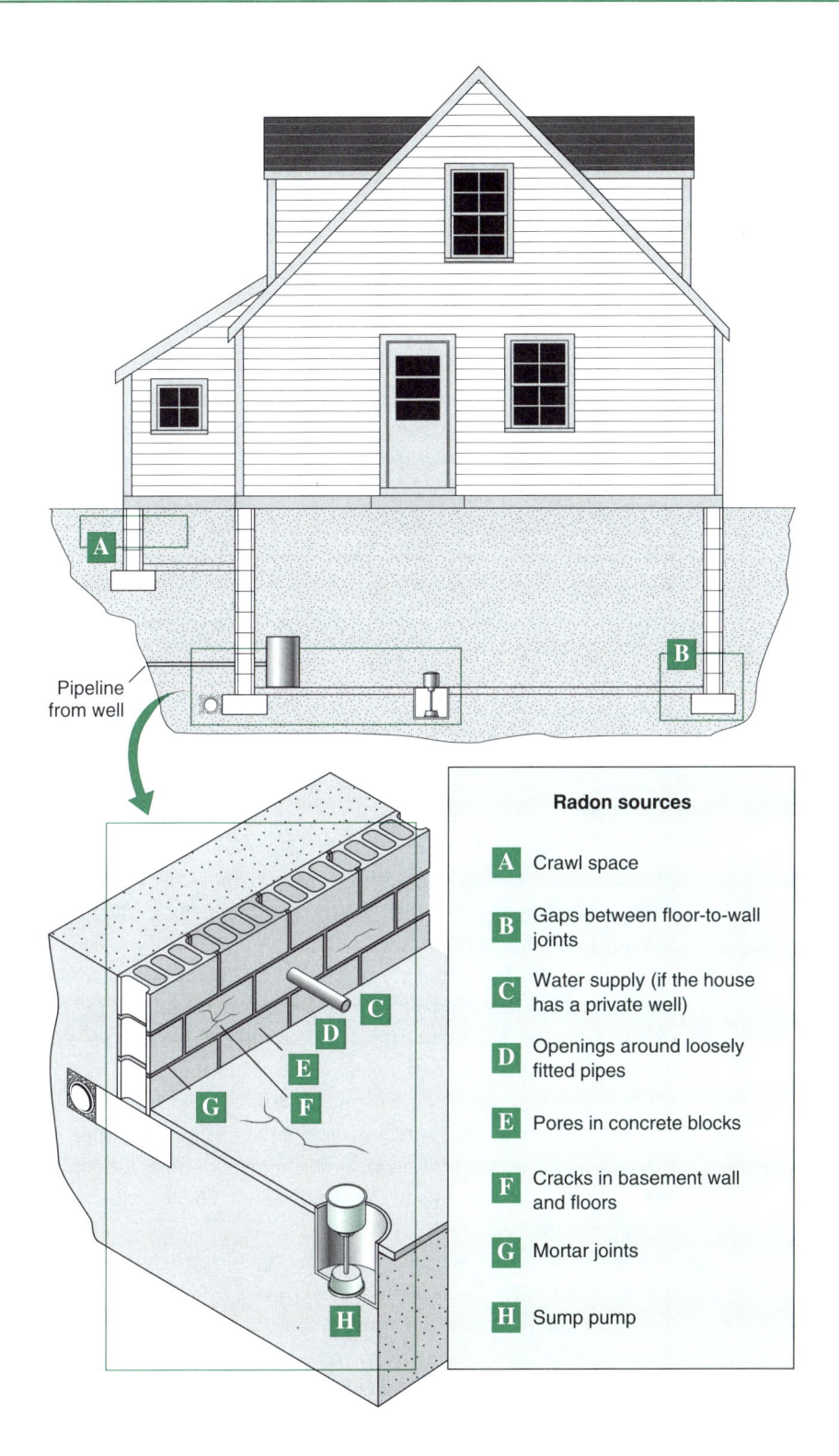

C Water supply (if the house has a private well)

D Openings around loosely fitted pipes

E Pores in concrete blocks

F Cracks in basement wall and floors

G Mortar joints

H Sump pump

- Openings around loosely fitted pipes
- Mortar joints

Although radon can also enter a home in well water, test the air in your home first. If you have high levels, and if you have a well, then check your water for radon.

Also, keep in mind that a neighbor's test results are not a good indication of whether your home has a radon problem. Radon levels can vary widely from home to home.

Short-term and long-term testing

There are two basic types of radon detectors: short-term units and long-term units. Short-term detectors, such as charcoal canisters, measure radon levels over a two- to seven-day period. Long-term detectors, such as alpha-track detectors, measure radon levels for three months to a year.

Radon levels within a home fluctuate with the weather, so short-term tests can be deceiving. A short, three-day test might catch the radon at an unusually high or low point, causing either unnecessary alarm or a false sense of security. In fact, a short-term test can vary from a long-term test by 200 to 300 percent.

If you select a short-term detector, follow "closed-house conditions" during the testing period. This means keeping all doors and windows shut except for normal entry and exit; in addition, turn off all fans.

How to test

Testing for radon is both simple and cheap, if you do it yourself. You can purchase radon testing kits from a local hardware store or other retail outlet. Their price usually includes the cost of having a laboratory analyze the detector and report the findings.

Whatever kind of detector you choose, make sure the manufacturer has successfully completed the EPA Radon Measurement Proficiency (RMP) Program. Most companies indicate on the package whether their test kit meets EPA requirements. In addition, state radon offices have a list of radon measurement companies that meet their state requirements.

The best time to test for radon, especially when using a short-term detector, is during the heating season, when your home is closed. However, you can conduct tests in the summer if you have central air conditioning and the house is otherwise closed.

Place the detector in a central location in the lowest living area of the house. Never place a detector in a crawl space or on the floor right next to a sump pump because a reading in those spots would have no relation to radon levels in the living area.

Radon detection kits should come with thorough instructions on how to set up the test. Be sure to indicate the dates that you begin and complete the test. Also, mail the detector to the lab *immediately* after completing the test. Failure to do so may result in a failed test.

Radon concentrations are typically reported in "picoCuries per liter of air (pCi/L)," with readings of 4 pCi/L or more being the EPA level of concern.*

To determine the concentration in your home, follow these testing steps:

Step 1. Do a short-term or long-term test (the U.S. EPA recommends starting with a short-term test). If your long-term test exceeds 4 pCi/L, that is sufficient reason to take action to reduce radon levels in your home. If you conduct a *short-term* test and the result is 4 pCi/L or higher, you should do a follow-up test (Step 2) to be sure.

Step 2. Follow up with either a long-term test or, if you need quick results, a second short-term test.

Step 3. Take action to solve the problem if the follow-up test was 4 pCi/L or more.

Radon and home sales

Many buyers want to know a home's radon level before purchasing the house, and some states now require homeowners to disclose test results over 4 pCi/L. But because real estate transactions can happen fast, you may not be able to make a long-term test. In such a case, you have several options:

- Perform two short-term tests, one immediately after the other. Then average the results.
- Perform two short-term tests at the same time and average the results.
- Hire a radon measurement company to perform a single forty-eight hour test with a continuous monitoring device that meets and is used in accordance with U.S. EPA requirements.

Another option is to test for radon *now* rather than wait for the question to arise during a transaction. One other point: if you are buying a new home, ask the owner or builder if it has radon-resistant features.

Radon-reduction methods

The techniques for reducing radon levels are outlined in Chapter 43.

Prepared with Henry Spies, UI School of Architecture—Building Research Council, and reviewed by the Illinois Department of Nuclear Safety, radon section.

---

*Some radon tests express the results as "working levels" (WL); 0.02 working levels is the equivalent of 4 pCi/L.

43 Reduce radon levels

Radon-reduction methods

A home in New Jersey had one of the highest residential radon levels ever recorded—3,500 picoCuries per liter (pCi/L), reported the October 1989 issue of *Consumer Reports*. But one day's work and about $1,300 brought levels down to 2 pCi/L, demonstrating that even high levels of radon can be brought within acceptable levels.

The following procedures have been tested successfully—by the Environmental Protection Agency (EPA), other research groups, or both—on homes with high indoor radon levels. If your home has a particularly high radon reading, you may need to use more than one strategy to bring radon within acceptable levels. Also, most radon-reduction remedies require the skilled services of a professional contractor who has experience in these procedures, has completed the U.S. EPA's "radon contractor proficiency" course, and has passed a comprehensive exam in radon mitigation.

Sub-slab suction

How it works. The lowest floor of most houses, other than those built over crawl spaces, consists of a concrete slab poured over the earth or on top of crushed rock (called "aggregate"). Sub-slab suction draws out radon, which accumulates under the slab, and vents it away from the house. It is most effective with foundations built on good aggregate or highly permeable soil.

Limitations. In some cases, it may be difficult to create enough suction to prevent radon from penetrating hollow-block basement walls.

Drain-tile suction

How it works. Some homeowners drain water away from the foundation of their house by perforated pipes called footing tiles. If these drain tiles form a continuous loop around the house and drain into a sump, you can apply suction to the sump. The suction pulls radon from the surrounding soil and vents it away from the house.

Limitations. This system only applies to houses with drain tiles that connect in a continuous loop around the home and that drain into a sump.

Heat recovery ventilation

How it works. A heat recovery ventilation (HRV) system increases house ventilation and uses the heated or cooled air being exhausted to warm or cool the incoming air. HRV systems can be designed to ventilate all or part of the house, but they are more effective in reducing radon levels when used to ventilate only the basement. If properly balanced and maintained, they ensure a constant degree of ventilation year-round.

Limitations. Heat recovery ventilators can be installed in any type of house, but they are generally not used in crawl spaces. Although energy losses are reduced when compared to other ventilation systems, heating and cooling bills will still increase.

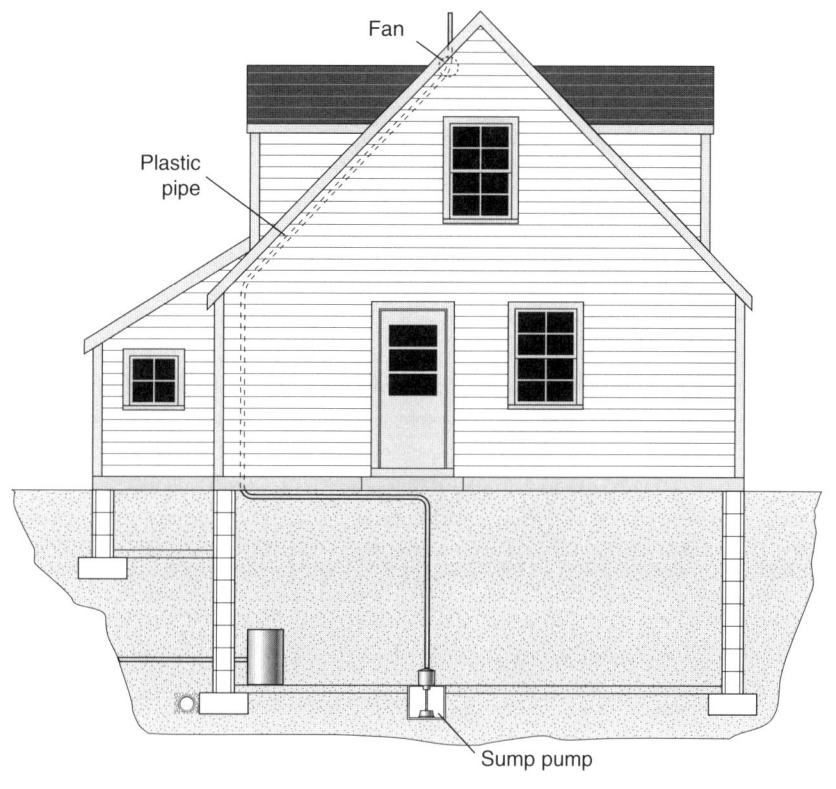

Sub-slab suction

Sub-slab suction is one of the most common and effective systems to control radon. With sub-slab suction, contractors run a 4-inch plastic pipe from either a hole in the floor or a sump (if one exists). The pipe runs up through the roof, while a fan inside the pipe sucks air (and radon gas) from the soil beneath the floor. The pipe acts like an oversized straw, drawing up the gas and venting it out through the roof to the outside air.

Covering exposed earth

How it works. By covering exposed earth with high-density polyethylene plastic, you reduce the flow of radon into the house. Exposed earth in basement cold rooms, storage areas, drain areas, sumps, and crawl spaces is often a major entry point for radon.

Limitations. Covering exposed earth may not be enough to solve your radon problem.

Sealing cracks and spaces

How it works. Sealing cracks reduces the flow of radon into your home and is often used with other radon methods. However, the U.S. EPA does not recommend the use of sealing alone to reduce radon; by itself, sealing has not been shown to lower radon levels significantly or consistently.

Limitations. It is difficult to find all of the cracks and gaps in your house. Also, settling and stresses can create more cracks, so continuing maintenance is necessary.

Block-wall ventilation

How it works. This system draws radon from the spaces within concrete block before it can enter the home. It is used most often with sub-slab suction.

Limitations. This system applies only to homes with hollow-block basement walls. Block-wall suction may not work if you cannot seal the top of the walls, the space between the walls and any exterior veneer, and openings that could be concealed by masonry fireplaces or chimneys.

Comparison of radon reduction methods

Method	Typical range of installation costs	Typical range of annual operating costs*	Typical radon reduction
Block-wall ventilation	$1,500 to $3,000	$150 to $300	50% to 99%
Covering exposed earth	$50 to $150 when polyethylene is used	None	Site-specific
Drain-tile suction	$800 to $1,700	$75 to $175	90% to 99%
Heat recovery ventilation	$1,200 to $2,500	$75 to $500 for continuous operation	25% to 50% if used for full house, 25% to 75% if used for basement
Sealing cracks and spaces	$100 to $2,000	None	0% to 50%
Sub-slab suction	$800 to $2,500	$75 to $175	80% to 90%

*Operating costs include fan electricity.
SOURCE: Based in part on information in *Consumer's Guide to Radon Reduction*, U.S. Environmental Protection Agency, August 1992.

Prepared with Henry Spies, UI School of Architecture—Building Research Council, and reviewed by the Illinois Department of Nuclear Safety, radon section.

44 Control dust, dust mites, and other allergens

Dust, asthma, and allergies

A growing body of scientific evidence, as well as patients' personal experiences, shows a strong link between indoor dust and illness, especially asthma and allergy.

Household dust consists of a wide array of particles that vary in size. The particles can come from both biological or nonbiological sources, and they can originate from outside or within the home. Among the most common culprits are dust mites, mold, mildew, fungi, algae, roaches and other insects, pollen, and animal dander.

In particular, a lot of attention has focused on the role of dust mites and cockroaches in human illness. Droppings from these insects, as well as their body parts, contain powerful allergens (any substance to which people can become allergic). Fine particles of these materials can become airborne as "dust," and can be drawn into the lungs, causing serious health problems.

A microscopic dust mite

Dust mites are microscopic members of the eight-legged Arachnid family and are related to spiders. They thrive in dust and high humidity and live primarily in mattresses, carpets, and upholstered furniture. In the United States, mite numbers seem to peak in July and August and persist at high levels through December. The lowest mite levels occur in April and May.

To deal with mites and other dust problems, there are two basic strategies: removal and avoidance.

Dust removal

Typical furnace filters are not designed to trap the fine particles and spores that cause health problems. Therefore, consider these alternatives:

Electrostatic precipitators. This electronic air cleaner, which offers high efficiency in removing fine particles, is installed in the return air duct near the home's furnace. Units have a removable dust-collecting plate that needs to be cleaned regularly.

Pleated filters. These filters are much more efficient than standard fibrous furnace filters; smaller ones do not require modifications to the furnace or ductwork. However, they are less efficient than electrostatic cleaners. Pleated filters should be inspected regularly and may need more frequent replacement than conventional filters.

Passive electrostatic filters. These filters are naturally charged by static electricity (they use no power) and fit into a standard filter slot. They provide fairly good filtration, but not as good as some pleated filters. Although somewhat expensive, these filters can be washed and reused.

Room-sized air cleaners. These cleaners are usually placed in sleeping rooms or other rooms where occupants spend most of their time. They can be moved to other rooms, but the most powerful units are consoles and are awkward to move.

Room-sized air cleaners have limitations, so their use must be coupled with other dust-control strategies. The most effective room-sized units are of either the electrostatic precipitator or HEPA type. An HEPA filter is a high-efficiency particulate-arresting filter, capable of removing most

One of the simplest and cheapest ways to reduce dust levels in your home is to switch from a standard filter to a pleated filter like the one shown above. The ridges of a pleated filter provide more surface space and, therefore, capture more dust than a standard, flat-surfaced filter.

of the particles that cause health concerns. Small tabletop units are not likely to be very effective, especially in large or open rooms.

Vacuum cleaners. If you use a vacuum cleaner for dust control, ideally it should have an HEPA filter. Filters in conventional vacuums are unable to trap fine particulates, ejecting them into the air. Also, vacuums that direct the exhaust stream through a container of water can disperse dust particulates in tiny water droplets. However, some high-performance conventional vacuums can perform almost as well as vacuums with an HEPA filter.

Avoid dry sweeping, which often stirs up more dust than it removes. If the dust-sensitive person must do the cleaning, he or she should wear a dust mask.

Avoidance

Avoiding exposure to dust is often more effective than dust removal, depending on the type of particulate involved. Basic knowledge of how and why dust particulates collect in the home is often necessary to prevent them from building up to unhealthy levels. Some examples follow:

Dust mites. Dust mites grow mainly in mattresses and bedding. But you can reduce their potential for trouble by washing all bedding and clothing brought into the bedroom every week to 10 days in the hot cycle of your washer. In addition, use mattress covers and pillow cases designed to block mites. These covers and pillow cases are available at surgical suppliers and health care outlets. Inexpensive plastic covers may also be effective.

Mites grow best at 75 to 80 percent relative humidity. In warm weather, air-conditioning can be used to maintain a humidity level of 50 percent or lower. You can use a humidity gauge to determine the level of humidity in the bedroom. If you operate a vaporizer or humidifier in the winter, these gauges can help you to avoid overhumidification. Try to keep indoor humidity levels in the range of 35 to 50 percent during winter—the optimum range for health.

Mold, mildew, fungi, and algae. These organisms require high moisture levels, wet surfaces, or standing water to grow. Therefore, control indoor moisture levels. Make sure drip trays under refrigerators and air conditioners are properly drained or regularly emptied. Eliminate standing water in the house, repair water damage, and kill visible growths on walls or other surfaces.

Roaches and other insects. Eliminate insect food sources, clean all greasy surfaces, reduce humidity levels, eliminate insect entryways to the house, and eliminate any areas of wetness (for example, leakage or condensation on plumbing).

> ✓ **CHECK IT OUT**
> For information on less toxic ways to control indoor insects: page 242.

Pollen. In the summer, you can reduce the entry of pollen and other outdoor pollutants by keeping the house closed and air-conditioned. Air-conditioning can also lower indoor humidity, controlling other organisms.

Animal dander. Dander, particularly from cats, causes serious problems for many people. Allergy testing can confirm whether this is true for you. If a pet cannot be removed from the environment, at least keep it out of sleeping areas.

Ambitious steps

A lot of the steps mentioned above make good sense for all homeowners, even if you do not believe you are particularly sensitive to dust. However, if your doctor has diagnosed that you have a dust allergy, he or she may recommend some even more ambitious steps to avoid exposure to dust—especially in the bedroom.

For example, your doctor may recommend:

Clean regularly, thoroughly. Clean the bedroom daily with a moist or oiled dust cloth, and do a more thorough job once a week.

Keep items that add to dust or are hard to clean out of the bedroom. Your allergist will probably recommend removing the following from the bedroom: carpeting, upholstered furniture, chenille bedspreads, hard-to-clean blinds, heavy curtains, pennants, wall hangings, dried flower arrangements, stuffed animals, and straw baskets. Washable shades or curtains and washable scatter rugs are preferred. Also, choose pillows stuffed with dacron or other synthetic materials, rather than feather pillows.

Don't forget the closet. Keep your closets as dust-free as possible. Anything stored in closets should be encased in storage bags or boxes.

Although these steps are most critical in the bedroom, dust-sensitive people can increase their comfort by observing them throughout the house.

Prepared with Joseph Ponessa, housing and energy specialist, Rutgers Cooperative Extension.

MAKE SURE HUMIDIFIERS DO MORE GOOD THAN HARM

Humidifiers can cause more problems than they solve if they are not cleaned regularly.

During the heating season, indoor air can become so dry that it aggravates cold, flu, and allergy symptoms. Humidifiers often solve this problem, adding needed moisture to the air. However, they also are excellent breeding grounds for mold, algae, and other organisms.

More importantly, some types of humidifiers—ultrasonic and cool-mist (spinning disk) humidifiers—can disperse mold particles, algae, and other organisms into the air. Fine particles contained in the tiny droplets of water that humidifiers emit can be drawn into the lungs. In addition, humidifiers can disperse the minerals from tap water into the air, so the use of distilled water (demineralized water) is often recommended.

The elderly, young, and those with breathing problems are the most sensitive to such problems.

All of this does not mean you should avoid using humidifiers; it means you should clean them regularly. Small humidifiers should be emptied, wiped clean, and refilled with fresh water daily. Sanitize them every two to three days. But be sure to unplug the unit before performing these tasks.

Sanitizing usually means rinsing with a dilute laundry bleach. Some ultrasonic humidifiers will be damaged by bleach, so consult the manufacturer's product literature for instructions and information on special products designed to cleanse this equipment. Three-percent hydrogen peroxide is recommended for sanitizing some humidifiers.

Prepared with Joseph Ponessa, housing and energy specialist, Rutgers Cooperative Extension; and Mel Bromberg, drinking water and health specialist, UI Cooperative Extension Service.

45 Reduce pollution from combustion equipment

Where there's smoke...

When any fuel is burned, the dangers from flames are obvious. But the hazards from smoke and other combustion products are often overlooked or underestimated.

With many fuel-burning appliances, smoke is exhausted to the outside through a flue. But some combustion appliances, such as gas ranges and unvented space heaters, discharge combustion products directly into the living area.

What are these combustion products, and how dangerous are they? Common household fuels—coal, wood, gas, oil, and kerosene—consist mainly of a group of chemicals called hydrocarbons. Under ideal conditions of temperature, air supply, and other factors, hydrocarbons react with oxygen and break down into carbon dioxide and water. This process also releases energy in the form of heat.

But conditions are rarely ideal. Contaminants are often present in the fuel, and combustion is rarely complete. Combustion byproducts can include strong irritants such as sulfur dioxide and oxides of nitrogen; unburned fuel; carbon (soot); carbon monoxide; formaldehyde; PAHs (polyclyclic aromatic hydrocarbons); and fine suspended particles (smoke).

These products are harmful, but the severity of problems they cause depends on their concentration in the air and the length of time you are exposed to them. For example, low concentrations of carbon monoxide can cause headaches, weakness, dizziness, and nausea; greater concentrations can kill. Nearly four hundred Americans die in their homes each year as a result of this invisible, odorless gas.

Oxides of nitrogen irritate the eyes, nose, and throat, and prolonged exposure can cause drowsiness. Long-term exposures to PAHs may increase the risk of cancer.

Before you get alarmed, keep in mind that some simple precautions and attention to routine maintenance will minimize the chances that indoor pollution problems will become serious. Observe the following precautions with various combustion equipment.

Furnaces and boilers

Problems arise when the chimney or flue becomes blocked, so inspect the chimney at the beginning of each heating season. If the chimney is relatively straight, the inspection can be done by inserting a small mirror in the chimney's clean-out door (first make sure the heater is turned off). Look for a clear path to the top of the chimney.

Also, inspect the metal flue pipe that connects the appliance to the chimney. Look for rust, holes, or weak spots. These pipes eventually corrode, and if they collapse, the house can rapidly fill with deadly gas. Another potential problem is that other equipment in the house may cause chimneys to "back-draft" (For details, see the accompanying sidebar, "Double messages and back drafts.")

Combustion equipment needs an adequate air supply, so furnace rooms should *not* be too tightly sealed. If the furnace room has a door, a 1- to 2-inch undercut along the bottom can help to provide needed air.

Note: some modern heaters are "direct-vented," which means combustion products are exhausted through a pipe in a nearby wall or window. No chimney is used, so some of these recommendations do not apply. Some units may also pump air for combustion directly to the furnace from outdoors; this ensures an adequate air supply.

Fireplaces and wood stoves

Homes with fireplaces and wood-burning stoves tend to have high indoor levels of combustion products, some of which are drawn in from the outside. Pollutants include fine particles, which can irritate the throat and lungs, and PAHs. What this means is that the familiar, "natural" odor of a fireplace or wood stove is actually loaded with harmful ingredients.

Avoid slow-burning, smoldering fires, which produce the largest amounts of these pollutants. Seasoned wood burns cleaner than green wood. Also, wood stoves sold after 1990 are required to meet federal emission standards, so they should be more efficient as well as cleaner-burning.

In addition to air-quality hazards, wood stoves can pose a fire hazard, usually because of improper installation. One of the most common mistakes is to disregard the specified clearance between combustible surfaces and the stove or its flue pipe.

Note: never burn wood treated with preservatives. As an example of what could happen, toxic metals that are safely bonded in CCA-treated wood are released by burning. Wood painted with lead-based paints also releases hazardous materials when burned.

Unvented fuel-fired space heaters

Unvented fuel-fired space heaters are intended for use indoors without directly expelling combustion products outside. Whether or not these combustion products cause problems depends on your health and how you use the appliances. The elderly, very young, and those who are sick (especially with breathing problems) are most likely to be harmed.

Do not operate a fuel-fired space heater in an enclosed space, such as a closed room. Closing the room increases pollution levels. In addition, it's important to follow the manufacturer's instructions. For instance, if you disregard kerosene heater instructions by setting the burner to produce a very high or very low flame, the heater may produce considerably more pollutants.

It's also important to use the proper fuel in kerosene heaters—low-sulfur, K1 fuel. Even more important, use only kerosene. Serious fires can occur if you use the wrong fuel—such as gasoline—either deliberately or accidentally. To avoid mistakes, fill and label fuel containers properly.

Gas-fired space heaters can produce pollutants as well, and problems can arise after prolonged operation in an enclosed space. As the flame depletes oxygen in the atmosphere, the combustion process is impaired, and production of carbon monoxide increases. Make sure gas heaters are equipped with an automatic shutoff that reacts to such conditions.

Gas ovens and ranges

Gas ovens and ranges produce pollutants, which can be vented outside by an exhaust fan. If not exhausted, household concentrations of these combustion products are usually harmless.

However, problems will develop if you use these appliances for supplemental heat in the home. Operation for many hours, day after day, can cause unacceptably high levels of pollutants, such as formaldehyde, carbon monoxide, oxides of nitrogen, and particulates.

Common sense

Household fuel-burning equipment is designed according to strict standards. But safety features and careful design cannot take the place of common sense. Use equipment according to the manufacturer's instructions. Also, regularly check these appliances for proper operation, and arrange for regular servicing and maintenance.

Prepared with Joseph Ponessa, housing and energy specialist, Rutgers Cooperative Extension.

CARBON MONOXIDE Q & A

What is carbon monoxide?

Carbon monoxide, also known as CO, is a colorless, odorless gas produced when most fuels are burned—fuels such as natural gas, propane, kerosene, charcoal, gasoline, and diesel fuel. It is produced in large amounts if an engine is not properly adjusted or if there is not enough oxygen where the fuel is being burned.

Why is carbon monoxide a problem?

Carbon monoxide is a highly poisonous gas. It binds strongly to hemoglobin in the blood, blocking the blood's ability to transport oxygen.

Repeated exposure to very small amounts of carbon monoxide over a period of hours or days can have a cumulative effect. Low doses can cause nausea, dizziness, and headache symptoms, which are frequently mistaken for the flu. Higher exposures can cause convulsions, coma, and death.

Because this gas is odorless, people may be unaware of its presence. Each year, about eight hundred deaths in the United States are attributed to carbon monoxide poisoning; about half of these occur in the home.

What sort of conditions produce a carbon monoxide problem?

In general, two situations can lead to carbon monoxide problems: malfunction of combustion equipment or the indoor use of combustion equipment intended for outdoor use.

Here are a few specific examples:

Blocked flue. Debris, nests, or the deterioration of the chimney flue can cause carbon monoxide to back up or leak into the house. A rusted pipe connecting the furnace to the chimney can also cause CO to leak into the house.

Back drafting. Under certain conditions, some combustion equipment can draw contaminants from the chimney flue into the house—a problem called "back drafting." (For more details on back drafting, see page 226.)

Insufficient air flow, improper burner adjustment. If the furnace is operating in a tightly sealed room, the combustion process will deplete the oxygen supply, increasing the production of carbon monoxide.

Faulty furnace heat exchanger. This malfunction allows combustion gases to mix with air in the ductwork. Frequent inspections are important, especially with older furnaces.

Fuel-fired space heaters. These units, both vented and unvented, are susceptible to some of the problems already mentioned. Check them on a regular basis, and carefully follow operating instructions. A yellow flame is an indication of low oxygen and high carbon monoxide production.

Never operate unvented fuel-fired space heaters in a closed room. Leave the door open.

Outdoor equipment used indoors. It is extremely dangerous to use fuel-fired cookstoves, outdoor space heaters, and charcoal grills indoors or in enclosed areas. Doing so can be a deadly mistake.

Are carbon monoxide detectors effective?

Several different types of carbon monoxide detectors are available. Inexpensive (about $5) detector cards have a treated spot that should change color when exposed to excessive levels of carbon monoxide. More expensive devices, which resemble smoke detectors, detect CO and emit an alarm; these detectors are powered either by batteries or house current.

The spot detector cards are unreliable, according to a *Consumer Reports* evaluation in May 1994. Also, the lack of an audible alarm is a major drawback because most carbon monoxide deaths occur when people are sleeping. Therefore, only battery-powered or plug-in detectors are recommended. But keep in mind that the battery-powered detector has a sensor unit that needs to be replaced every two to three years. The sensor unit costs about $20.

Prepared with Joseph Ponessa, housing and energy specialist, Rutgers Cooperative Extension.

DOUBLE MESSAGES AND BACK DRAFTS

During the 1970s and into the '80s, there was a major push for homeowners to tighten their homes to save energy. But in the mid-1980s, attention began to shift to radon and other indoor air-quality problems, with a stress on ventilation.

Consequently, many consumers now want to know: Is there a direct conflict between sealing air leaks for energy conservation and providing ventilation to improve indoor air quality?

Not necessarily.

"Old" homes, including many built prior to the first energy crisis of the 1970s, were usually built without much concern about building tightness. As a result, most of these structures have many air leaks and are well ventilated. Even if a house has been sealed tightly to prevent energy losses, there should still be plenty of entry points for fresh air.

However, occupants of homes that have been tightened to save energy should be aware that combustion appliances need fresh air, or "makeup air," to operate efficiently. Without adequate makeup air, combustion becomes "starved," producing excessive soot and carbon monoxide (which is poisonous).

In a tightly sealed house, another problem is that by operating a powerful ventilation device, such as a surface-mounted exhaust system on a range top, you can create a "back draft" in the flues of other combustion devices. A back draft can pull exhaust fumes back into the house, drawing poisonous gases into the home.

A roaring fire in the fireplace can also cause a back draft in the flues of other heating equipment (such as the furnace), especially if the fireplace does not have glass doors.

The problems of back drafts and inadequate makeup air can be easily prevented by opening a window slightly (about ½ inch) when operating exhaust fans or fireplaces. During the heating season, leave an exterior vent or window open slightly in the furnace room, but be sure to protect the area from freezing temperatures.

Some contractors are well aware of the need to balance energy efficiency with adequate ventilation. So, when constructing a new house, determine what the contractor is doing to find this optimum balance.

Prepared with Joseph Ponessa, housing and energy specialist, Rutgers Cooperative Extension.

46 Prevent contamination by formaldehyde and asbestos

Signs and symptoms of formaldehyde

The pungent odor of formaldehyde that you may remember from high school biology lab can turn up in the home as an irritating gas. It is emitted from certain products, including certain types of plywood and paneling, furniture, and some foamed-in-place insulation.

Formaldehyde can cause watery eyes, burning sensations in the eyes and throat, nausea, and difficulty in breathing in some people. But it is important to note that these symptoms can also be caused by many other pollutants.

Although some research suggests that prolonged exposure to high levels of formaldehyde may cause cancer, studies of workers exposed to formaldehyde have failed to provide consistent support of this.

Sources of formaldehyde

Many of the products that use formaldehyde contain very small amounts and may cause problems only for the few people who have extreme sensitivities. More important are those products that (a) contain substantial amounts of formaldehyde and (b) are used in large quantities throughout the house.

The following products, which contain a particular type of adhesive—urea formaldehyde—are of special concern:

Major sources of formaldehyde

Material	Use
Hardwood plywood paneling	Paneling
Medium-density fiber board	Veneered or laminated furniture, drawer fronts
Particle board	Subflooring, laminated countertop substrate

Another source of formaldehyde has been urea formaldehyde foam insulation (UFFI), a foamed-in-place insulation material. This product, pumped into wall cavities in the 1970s and early 1980s, was eventually banned. Although the ban was later lifted, the product has been rarely used since. If urea formaldehyde foam insulation was installed in your home five to ten years ago—or more—it is unlikely to still be releasing formaldehyde.

Detecting formaldehyde

It is sometimes possible to detect a formaldehyde problem simply by its characteristic odor. Another option is to test for its presence in the air by hiring an environmental testing firm, listed in the telephone directory.

Do-it-yourself kits usually involve placing a detector in a room for a couple of days and sending it to a laboratory for analysis. Although the accuracy of these tests is questionable for very low concentrations, a positive result can confirm a problem.

Formaldehyde can be found in nearly all homes, at least in low concentrations. Those homes with few sources of formaldehyde may have levels of about 0.02 to 0.07 parts per million (ppm), whereas homes with particle board subflooring may have concentrations of 0.06 to 0.15 ppm. In the past, many of the most serious problems have been associated with mobile homes, but mobile-home manufacturers have since made improvements.

Research suggests that some health problems can occur at levels of 0.10 ppm, with sensitive persons affected at levels of 0.05 ppm. Consult a knowledgeable physician if you suspect you are reacting to formaldehyde exposure.

How to reduce formaldehyde exposure

- Block formaldehyde emissions by covering exposed particle board with two coats of polyurethane varnish or, preferably, lacquer.
- Use substitutes for particle board. Some examples include waferboard and softwood interior-exterior plywood.
- Use solid-wood furniture instead of veneered furniture. Also, use solid wood for cabinets.
- Use drywall instead of wall paneling.
- Use air-conditioning and dehumidifiers to maintain moderate temperatures and reduce humidity levels. Heat and high humidity increase the rate at which formaldehyde is released.
- Increase ventilation, particularly after bringing new sources of formaldehyde into the house.
- Wash permanent-press fabrics before use. (Some fabrics are treated with formaldehyde to make them resistant to wrinkling.)

Asbestos in your home

Asbestos, a mineral fiber often found in the home, is another material that can become an air pollutant. But the mere presence of asbestos in your home is not hazardous. Asbestos becomes dangerous only when the material is damaged and asbestos fibers are released into the air. Breathing high levels of asbestos fibers over a long period can lead to an increased risk of lung cancer and other respiratory diseases.

Sources of asbestos

Until the 1970s, when the hazard came to public attention, asbestos was added to a variety of products such as insulation and asbestos cement roofing shingles and siding to strengthen them and provide heat insulation and fire resistance.

Most products made today do not contain asbestos, and the few products that still contain asbestos that can be inhaled are required to be labeled as such. Common products that might have contained asbestos in the past (and may be present in some homes today) include:

- Steam pipes, boilers, and furnace ducts insulated with an asbestos blanket or asbestos paper tape. They may release fibers if damaged, repaired, or removed improperly.
- Resilient floor tiles (vinyl, asbestos, asphalt, and rubber), the backing on vinyl sheet flooring, and adhesives used for installing floor tile. Scraping or sanding may release fibers.
- Cement sheet, millboard, and paper used as insulation around furnaces and wood-burning stoves. Cutting, tearing, and removing this insulation may release fibers.
- Door gaskets in furnaces and wood and coal stoves. Worn seals can release fibers during use.
- Soundproofing or decorative material sprayed on walls and ceilings. Loose, crumbly, or water-damaged material may release fibers.
- Patching and joint compounds for walls and ceilings and textured paints. Sanding and scraping will dislodge fibers.
- Asbestos cement roofing, shingles, and siding. They are not likely to release fibers unless sawed or cut.
- Older household products such as fireproof gloves, stove-top pads, ironing board covers, certain hair dryers, and artificial ashes and embers for gas-fired fireplaces. Worn items will release fibers.
- Some automobile brake pads and linings, clutch facings, and gaskets. Normal wear will release fibers.

What to do about asbestos in the home

Don't panic if you think asbestos may be in your home. Usually, it's best to leave asbestos material alone if it is in good condition. Disturbing it may create a health hazard where none existed before. If the asbestos is damaged, seek expert help to repair or remove the material.

Repair involves sealing, encapsulating, or enclosing the material to prevent the asbestos fibers from being released. Exposed insulation on hot-water or steam pipes could be covered with a protective wrap or jacket.

Removal usually is the most expensive route to go. Removal also poses the greatest risk of releasing asbestos fibers, so this should be the last option considered in most situations. However, it may be your only option when remodeling or making major changes in your home.

Contacting a professional

Because of the hazards involved, asbestos repair and removal should be done by professionals who have received special training. Ask for proof of training and licensing in asbestos work, such as completion of a training course approved by the U.S. Environmental Protection Agency (EPA).

Professionals are also available to inspect your home for asbestos. State and local health departments or EPA regional offices may have listings of licensed professionals in your area who can handle inspections and removal. If asbestos must be removed as part of roofing work, another option is to call (800)USA-ROOF for names of qualified roofing contractors in your area.

Prepared with Joseph Ponessa, housing and energy specialist, Rutgers Cooperative Extension.

Seattle

Seattle residents put their homes to the test

People are healing their homes in Seattle, Washington.

They are using the Home Environmental Assessment List (HEAL) to pinpoint indoor pollution problems. Then they are turning to a set of low-cost, and sometimes no-cost, solutions.

It's all part of the Master Home Environmentalist Program, an innovative creation of the Metrocenter YMCA, the Washington Toxics Coalition, and the Home Toxics Task Force. Volunteers go into homes with a HEAL survey to evaluate a home's environmental quality, says Philip Dickey of the Washington Toxics Coalition. In addition, they offer a do-it-yourself form that homeowners can use themselves.

According to Dickey, survey questions focus on the building structure, nearby environment, dust and lead control, moisture problems, indoor air, and hazardous household products.

"People are often surprised by the number of hazardous products they have in their homes," Dickey says. "There are lots of instances where people move into a home and inherit certain hazards. Or people who live in a home a long time tend to accumulate a lot of things—some of which are hazardous."

The HEAL survey ranks answers according to the level of concern—lower, medium, or higher. Then it directs residents to an "action step," a specific action people can take to solve the problem. (See the accompanying example.) When the Master Home Environmentalist Program was pilot-tested in 1993, participants discovered the greatest pollution problems in the dust and lead category. For example, Dickey says peeling paint (a high risk for lead dust) was reported in almost half of the sixty-six HEAL pilot surveys.

Moisture problems ranked second in seriousness, followed by problems with indoor air pollution and hazardous products. Among indoor air pollution, the most common risk was the presence of particle board, a common source of formaldehyde.

As a result of the 1993 survey, participants reaped measurable gains. Lead samples were collected at their homes before and after the program, and Dickey says they saw a 52-percent reduction in lead in dust.

"Lead in dust is the best predictor of blood-lead levels in children," he notes. "The magnitude of this reduction indicates that the program can have major impacts on the health of children in particular—and for other residents as well."

Home Environmental Assessment List (HEAL): An excerpt

The following are a few sample questions from the dust and lead section of Seattle's HEAL survey. Residents answer questions and are then directed to a numbered list of "action steps." The complete HEAL survey includes more than 90 questions and more than 80 action steps. For information about Seattle's Master Home Environmentalist Program, you can contact the Metrocenter YMCA, 909 Fourth Ave., Seattle, WA 98104.

Dust and lead control

	Level of concern			
	Lower	*Medium*	*Higher*	*Action step*
Is any paint peeling or flaking inside or outside the home?	No	Don't know	Yes	16
If yes, is the paint lead-based paint?	No	Don't know	Yes	13
What percentage of the living space is carpeted?	0 to 25%	25 to 75%	76 to 100%	20
How do you control track-in of dust or dirt?	Remove shoes	Doormat/ hall rug	I don't	14
Does every door have a door mat?	Yes	Some	None do	14

How often do you:	4 times per month	1 to 3 times per month	Less than once per month	
Vacuum home				23
Dust				23
Scrub bathrooms				23
Shake rugs/door mats				23
Wash beddings/sheets with hot water				23
Air out home/open windows				29

Home Environmental Assessment List (HEAL), cont.

Dust and lead control

Level of concern

	Lower	Medium	Higher	Action step
If anyone in the home has allergies, do you have plastic covers on your pillows and/or mattresses?	Yes	—	No	28
How do you clean area rugs?	Send out	Vacuum both sides	Vacuum surface	24
If you have a furnace, when was the ductwork last cleaned?	0 to 1 year ago	1 to 4 years ago	More than 4 years ago	32
How often do you change the furnace filters?	More than once per year	Every 1 to 2 years	Every 2 years or more	33
Does the furnace have an outside combustion air supply?	Yes/not applicable	Don't know	No	34

Action steps for excerpt above

13. Seek expert advice on reducing lead levels in the home.
14. Reduce track-in of dust and lead. Take off shoes or install high-quality doormats at all entrances.
16. Have paint tested for lead if home is pre-1978, and especially if there are children present.
20. Vacuum with an upright or powerhead vacuum.
23. Increase frequency of vacuuming or cleaning.
24. Vacuum both sides of area rugs every six months.
28. Consider covering pillows and/or mattresses with plastic.
29. Open windows more frequently and keep bedroom window open at night.
32. Have furnace ductwork cleaned regularly (about every five years).
33. Change furnace filters at least twice a year.
34. Have an outside combustion air supply installed for furnace.

47 Recognize that it's not always necessary to control indoor insects

Insects and aesthetics

In a cartoon from Gary Larsen's *Far Side* collection, two insects are relaxing and reading the newspaper in their underground home, and just beside them is a massive mound of eggs, ready to hatch. The insect wife turns to the husband and says, "You know, Vern...the thought of what this place is gonna look like in about a week just gives me the creeps."

A human couldn't have put it better. When we spot an insect inside our house, we sometimes cringe, even though most indoor insects do not pose serious threats to us, our homes, or our belongings. There are infamous exceptions, such as the termite, but most indoor insects are little more than an aesthetic or nuisance problem. This can include carpenter ants, which some homeowners fear will do structural damage to their home. It's true that carpenter ants will burrow into the damp wood of a home, but they typically affect only about a 2-foot stretch of a two-by-four—wood often already damaged by moisture. What's more, they usually do not spread beyond one or two nests in a single home.

With a lot of insects, the question is how many you can tolerate. For most people, the sight of one cockroach is the limit, especially considering the rate at which they breed. But what about other indoor insects? If an insect doesn't do damage, can you handle its presence in your home? Should you treat with pesticides? These are individual decisions.

The accompanying chart provides notes on several common insects. Understanding the damage that these insects can do may be the key to whether you try to control them.

If you decide that the potential for damage or the nuisance factor warrants taking action, read the chapters that follow. They provide tips on pesticide selection, as well as alternatives to traditional pesticides.

What should you ask a pest-control operator?

If you decide to contact a pest-control company for help, try to find one that offers an "integrated pest management" program, in which the company's pest-control operators use the least toxic control methods and turn to pesticides as a last resort. (Locating such companies may be difficult.) Also, find out if the operators will "spot-treat" for a problem. This means they treat only the critical areas, rather than make a blanket treatment over a large area. After pesticides are applied in your home, it makes good sense to open the windows and air out the house for as long as possible.

One of the most steady sources of income for pest-control companies is a contract in which they spray your house monthly to prevent problems. Unless you have an extremely low tolerance for the sight of any insects indoors, the regular monthly treatments offered by extermination companies are not a good idea. These precautionary treatments expose you to more pesticides than is usually necessary. Instead, treat for insects only if you discover an infestation and decide that action is necessary.

Common household pests

Residential pests

Pest	Appearance	Damage
Bean weevils	1/8", brown, adults with snouts	Feeds on grain products
Carpet beetle	1/4", brown, elongated larva	Creates holes in woolens and loosens carpet nap
Drain fly	1/8", triangle-shaped black fly	Larvae feed on debris in drainpipe
Flea	1/16" to 1/8", dark brown, jumps	Bites pets and humans, causes red bites
Flour beetle	1/8", reddish brown	Feeds in cereals, flour, grain products
German cockroach	1/2", fast-moving, brown, active at night	Nuisance in kitchen

Common household pests, cont.

Pest	Appearance	Damage
Head louse	1/16", light brown	Feeds on human blood on scalp
Indianmeal moth	1/4" brown-and-white moth, 1/2" white caterpillar	Larvae feed on grain products, nuts, chocolate, dried fruit
Oriental cockroach	1 1/4", fast-moving, dark brown, active at night	Nuisance around drains, especially in the basement
Silverfish	1/2", fast-moving, silvery, active at night	Nuisance in bathroom, basement

Invaders from outdoors

Pest	Appearance	Damage
Ant	1/8"; elongated; red, brown, or black	Nuisance
Boxelder bug	1/2", red and black	Nuisance indoors in winter and on outside walls in fall

Pest	Appearance	Damage
Centipede	1" or longer, many legs, fast-moving	Nuisance
Clover mite	Tiny, dark-red to orange specks	Nuisance, enters homes in early spring or late fall
Cluster fly	House fly-like	Nuisance indoors due to large numbers during winter
Earwig	5/8", brown with pincers at rear of body	Nuisance
Field cricket	1", black, jumps	Nuisance at night while chirping
House mouse	Grayish mouse with pointed head and long tail	Feeds on grain products, onions, potatoes, apples; fouls other products with urine and feces
Millipede	1", dark brown, many legs, slow-moving, coils when disturbed	Nuisance in basement

Common household pests, cont.

Pest	Appearance	Damage
Sowbug	¼", gray, armadillo-like	Nuisance in basement
Spider	8 legs, fast-moving	Mostly nuisance, some may bite
Wood roach	1¼", dark brown	Flies into home in spring, attracted to lights at night

Stinging insects

Pest	Appearance	Damage
Bald-faced hornet	1", black with cream face and tail, nest in tree is football-shaped and -sized (sometimes larger)	Stings if disturbed
Honey bee	½" to ¾", brown and black, may nest in wall void	Stings if disturbed
Paper wasp	1", reddish brown, elongated, umbrella nest under eave	Stings if disturbed

Pest	Appearance	Damage
Yellowjacket 	½" to ¾", yellow-and-black striped, nest underground or in wall void	Stings if disturbed, particularly in fall
Structural pests		
Carpenter ant 	¼" to ¾", elongated, black	Nuisance, hollows out damp wood for nest
Carpenter bee 	1", yellow and black	Bores ½"-diameter hole in wood
Powder-post beetle 	⅛", brown to black	Creates pinhead-sized holes in wood, reduces wood to powder, damage usually localized
Termite 	⅛" white, wingless worker; ¼" black, winged reproductive	Hollows out wood, but leaves no sawdust or external holes

Prepared with Philip Nixon, John Lloyd, and Rick Weinzierl, entomologists, UI Cooperative Extension Service.

48 Select the right household pesticide

Identify the problem

The first step in selecting a pesticide is to accurately identify the pest that needs to be controlled. If you don't know what it is, don't guess. Contact a local pest-control firm or the nearest Cooperative Extension Service office for assistance. Chapter 47 provides additional tips on detecting pest problems.

Once you have identified a pest and decided that using a pesticide is necessary, select an appropriate type of treatment. Choices include residual pesticides, baits, and space sprays. Remember to read the pesticide labels carefully to find out which products can be used for the treatment you have in mind.

Residual pesticides

Residual pesticides are sprayed on surfaces and allowed to dry; they are intended to work against pests for a few days or weeks after application. They remain toxic to insects and to pets and humans, so use them with care. Follow these general guidelines.

Cracks and crevices. The most effective and safest use for residual pesticides is to apply them in cracks and crevices where ants, roaches, and other insects hide. Key application sites are the gaps and spaces along baseboards and behind appliances, where humans and pets are unlikely to contact the residue.

Barriers. Some residual pesticides can be used to create chemical barriers. For example, spraying along a doorway threshold or around the foundation (including a 6-inch band of soil adjacent to the foundation) can keep out ants, crickets, earwigs, and spiders; spraying along windowsills can keep out clover mites. Before using a pesticide as a barrier, however, first determine whether the places where insects enter can be plugged with caulk or weatherstripping; this way, you may be able to avoid insecticides.

General surface sprays. It is *not* a good idea to use residual pesticides to spray surfaces throughout the interior of a house. Most of the pesticide will wind up on surfaces that the pests will not contact, but humans will.

Baits

Rodent baits. Rodent baits can be deadly, so place them only where they are inaccessible to children and pets. Always place them inside lockable bait boxes, which can be purchased where baits are sold. *Never* place a rodent bait in an open container under the sink. Traps, including those that do not kill the offending rodent, are alternatives to baits.

Roach and ant baits. Roach and ant baits use pesticides that are less toxic to humans than those in rodent baits. Some labels state that these baits are safe enough to place on countertops, but logic suggests this is a poor idea; a countertop usually is not the most effective location. You will have much more success in controlling roaches and ants if you place baits where the insects enter the room or emerge from cracks and voids. Good sites for bait containers are near pipes, corners, and damp areas.

Don't expect to see dead insects around the bait containers; slow-acting pesticides are used so that ants and roaches take the bait back to the nest, where they share food, poisoning others in the process.

Remember, place baits in areas where cats and dogs will not claw or chew them open and children will not find them and play with them.

Space sprays (aerosols)

These pesticides, usually aerosols, are sprayed or otherwise released into the air to control such pests as flies and mosquitoes. Many labels state that it is safe to spray these chemicals indoors and remain in the house, but common sense suggests that it is a good idea to leave the treated rooms or building for at least a few hours.

In general, space sprays are not very effective for the control of most pests. Additionally, tiny aerosol droplets of the pesticide settle on surfaces throughout the house—many of them surfaces where pests are not active. For example, the "bombs" used to eradicate fleas have limited effectiveness. See Chapter 49 for details on more effective approaches to flea control.

Prepared with Philip Nixon and Rick Weinzierl, entomologists, UI Cooperative Extension Service; and Christine Wagner-Hulme, UI graduate research assistant.

49 Control household pests with less toxic alternatives

Understanding pests

If you're looking for ways to avoid using synthetic pesticides in the home, there are a variety of strategies from which to choose. Keep in mind that insects and rodents have a reason for visiting your home—available food, water, and shelter. Also, their presence means they must be finding a way to get into the house. You can eliminate many problems if you emphasize these nonchemical control efforts:

- Exclusion: eliminating entry points to the house.
- Sanitation: eliminating sources of food, water, and shelter.
- Habitat modification: eliminating or disrupting the habitat where a pest resides.

In addition, you can try other nonchemical approaches, such as subjecting insects to extreme temperatures or setting traps, or you can rely on certain insecticidal dusts, botanical insecticides, and insecticidal soaps that are less toxic than traditional pesticides.

Exclusion

- Place screening over openings to keep out flies, mosquitoes, and beetles. Placing screens on air vents and ducts can also stop the movement of cockroaches.
- Use caulk to close cracks and crevices, especially around water pipes and windows. These are prime avenues of entry and travel for several common pests.
- Seal or repair openings on the exterior of the house to prevent bats, mice, bees, and wasps from either entering the house or building nests. Also, screen attic vents.
- Inspect plants and food products for infestations when you buy them and before you bring them into your house.

Sanitation

- Dispose of food waste in sealed plastic bags, or in garbage or compost containers that have snap-on lids. Empty garbage on a weekly basis or more often. Do not allow food to remain in the garbage disposal overnight.
- Store bulk food in sealed containers, either plastic containers with snap-on lids or glass jars with rubber gaskets.

- Keep pantry and cabinet shelves, toasters, and couch cushions "crumb-free."
- Discard overripe fruits, onions, and potatoes to control fruit flies and fungus beetles.
- Remove bird nests from your house. They can harbor beetles, clothes moths, mites, and lice.
- Vacuum regularly and thoroughly to reduce populations of fleas, carpet beetles, dust mites, and several ground-dwelling insects and insect relatives. Steam-cleaning carpets and upholstered furniture further reduces populations of common pests, especially fleas.
- Remove pet food and water bowls from the floor at night, especially if cockroaches are a problem.

Habitat modification

- Remove debris and fallen leaves near foundations to reduce earwig, cricket, spider, sowbug, millipede, and centipede populations.
- Remove limbs that touch the house.
- Store wood away from the house because it can harbor termites, carpenter ants, ground beetles, and spiders.
- Use a dehumidifier, especially in basements, to create a dry environment that reduces problems with sowbugs, centipedes, silverfish, firebrats, and dust mites.

Temperature control

You can eliminate pantry pests, clothes moths, and carpet beetles (as well as many other insects) by subjecting infested materials to extremely hot or cold temperatures.

In general, you can eliminate all developmental stages of these insects by placing infested materials in a household freezer for three days or by exposing them to temperatures above 140°F for at least thirty minutes. (For many problems, freezing is usually safer than heating, although neither fate will appeal to the flea-infested family cat.)

Mechanical control

To eliminate visible or less mobile insects, there's the old reliable rolled-up magazine, fly swatter, or the sole of your shoe. The vacuum cleaner is the most appropriate tool for eliminating boxelder bugs, black vine weevils, and elm leaf beetles that have moved indoors. Few instructions are necessary on this topic; this is just a simple reminder that certain old-fashioned methods still work.

Insecticidal dusts

All insects have a waxy cuticle covering on their bodies; it prevents excessive loss of water. Abrasive or sorptive dusts may damage or destroy this wax coating and kill the insect as a result of dehydration.

Abrasive and sorptive dusts, used as insecticides, tend to be less hazardous to humans than most synthetic pesticides because they are nonvolatile (do not become a vapor in the air) and pose little risk of injury when they contact skin. Even so, they can irritate mucous membranes and the respiratory tract, so use goggles and a dust mask to protect your eyes and lungs from dusts. Three of the most common insecticidal dusts are silica aerogel, diatomaceous earth, and boric acid.

Silica aerogel. This dust is particularly effective against insects in confined areas. Pest-control operators often use silica aerogels against ants, bees, boxelder bugs, cockroaches, crickets, silverfish, and spiders. When applying silica aerogels, leave a thin coating in cracks and crevices; treatments remain effective for several months if they remain dry.

Diatomaceous earth. If diatomaceous earth is ground to the correct particle size, it is effective as an insecticide because it abrades, or damages, the insect cuticle. Diatomaceous earth that is packaged for use in swimming pool filters and similar applications is *not* ground to the correct size to be used as an insecticide.

EPA-registered products that contain diatomaceous earth may be used in attics and wall voids to kill ants, roaches, and similar pests. Also, you can use diatomaceous earth to protect packages of seeds from insect damage. Although labels for some products include instructions for use outdoors on soil and plants, moisture greatly reduces the effectiveness of diatomaceous earth as an insecticide outdoors.

Boric acid. Boric acid is commonly used against cockroaches, fleas, silverfish, and similar pests. Apply it as a fine coating of dust in cracks and crevices or along the baseboards behind cabinets and heavy appliances—places where children and pets cannot contact it. Use only products that are manufactured and approved specifically for use as insecticides. *Be aware that swallowing boric acid, even in quantities of less than a tablespoon, can be fatal to small children.*

Botanical insecticides and insecticidal soaps

In addition to insecticidal dusts, two other alternatives to traditional pesticides include botanical insecticides and insecticidal soaps.

Botanical insecticides, sometimes referred to as "botanicals," are naturally occurring insecticides derived from plants. Insecticidal soaps are soaps that have been formulated specifically for their ability to control insects. Although these products are most commonly used to kill insects in your garden, they are also used to treat house plants. For more information, see Chapter 11.

Traps and baits

Traps use sticky substances or funnel-shaped openings to capture insects; they catch only the insects that wander into or onto the trap. With bait stations, on the other hand, the insect collects a poisoned bait, takes it back to the nest, and shares it with the rest of the colony. As a result, baits often do a better job than traps in controlling a large pest population.

> **CHECK IT OUT**
> For more information on baits: page 240.

Traps are more useful for *monitoring* a pest than for *reducing* a pest population—unless the population is isolated or confined to a small area. For instance, sticky cockroach traps are excellent tools for determining where infestations are located, but they do not control the roach population. However, there are some exceptions to this rule; see Chapter 13 for more details.

Prepared with Rick Weinzierl and Philip Nixon, entomologists, UI Cooperative Extension Service; and Christine Wagner-Hulme, UI graduate research assistant.

50 Select household cleaners wisely

Cleaners: How big a problem really?

In the world of cleaning, "phosphate" has been a dirty word for a long time among consumers concerned about the environment. Phosphate increases algae growth in water, reducing oxygen levels for fish. It also contains traces of arsenic.

Today, most manufacturers have gone to phosphate-free laundry detergents, although phosphate-free *dishwashing* detergents are not as common (read the label to find out about phosphates). What's more, today's phosphate-free detergents do a much better job than their predecessors in the 1970s.

So, if phosphate detergents are not the issue they once were, what concerns do household cleaners pose in the 1990s? The following are some good reasons why you still need to select cleaners carefully:

Indoor air quality. Research studies comparing indoor and outdoor pollution have yielded some surprising results: in nearly all homes tested, indoor pollution levels were consistently higher than outdoor levels, even in polluted areas. Specifically, the same group of about a dozen pollutants (volatile organic compounds, or VOCs) were seen in almost every home tested. These pollutants come mainly from products and materials in the house, such as cleaning agents, waxes, polishes, spot and stain removers, and even newly dry-cleaned garments.

Fortunately, all concentrations were at levels well below those permitted in the workplace. While there is no evidence that households are at risk from low concentrations of these pollutants, neither can we say there is no risk at all. The main concern is that very little is known about possible effects when low concentrations of the more hazardous materials mix in the air.

The heavy metal connection. Various household cleaners contain trace elements of heavy metals, such as cadmium, chromium, arsenic, copper, lead, mercury, nickel, silver, or zinc. Although household cleaning products are used in small quantities in individual homes, collectively they contribute substantially to the amount of heavy metals reaching wastewater-treatment plants. Heavy metals can contaminate sludge—the solid-waste byproduct created when water undergoes treatment. This makes sludge disposal a problem for treatment plants.

Several years ago, the Washington Toxics Coalition in Seattle analyzed phosphate-free laundry detergents for heavy metals. They found greater amounts of heavy metals in powdered laundry detergents than in liquid laundry detergents.

Safety. Household cleaners vary greatly in their levels of toxicity. But some, such as oven cleaners and drain cleaners, are quite hazardous if not handled properly.

What is a consumer to do?

There are about as many opinions on the environmental hazards of household cleaners as there are products on the shelves. But one thing is safe to say: water remains the most important and least toxic universal cleaning solvent. When combined with detergent, it is a very effective cleaner for food, salt, or sugar stains and oily residues.

For the tougher jobs that ordinary soap and water cannot handle, consider the following points:

- Select the least toxic alternative that will do the job.
- Read the label to find out what precautions are called for, and follow instructions carefully. If the product is to be diluted, do it properly.
- Don't overuse a product. Try to do the job with the least amount of product possible.
- To reduce the amount of heavy metals in wastewater, use liquid laundry detergents and liquid bleaches.
- To improve indoor air quality, use cedar panels instead of mothballs, and air out your dry cleaning outdoors, when possible. Recently dry-cleaned clothes can release volatile organic compounds (a broad group of chemicals) into the air, but in low concentrations.
- Look for products that have replaced petroleum distillates with safer citrus-based solvents. Many paint strippers, household cleaning concentrates, and mechanics' hand cleaners have gone to citrus-based solvents.

What about 'natural' cleaners?

Some people suggest that we return to using "natural" cleaners, such as vinegar, table salt, and lemon juice—cleaners that were used before today's commercial products were available. But it's important to realize that these products are also chemical compounds—natural chemicals, but chemicals nonetheless.

Alternative cleaning chemicals can be just as hazardous as commercial products if they are used inappropriately. They can be caustic to skin or irritating to the eyes, so store them safely out of the reach of children.

To be safe, *do not mix chemicals.* Experimental cleaning "recipes" can be dangerous to people, as well as harmful to fabrics, materials, and surfaces. The classic example is the mixture of chlorine bleach and ammonia, which creates a toxic gas.

If you decide you want to try other alternatives to commercial cleaning products, be sure you have some background information on them. Here are some of the most common alternatives:

Alcohol. Isopropyl and denatured alcohol are sometimes used to remove stains and mildew, but they can cause some dyes to fade. Do not use alcoholic beverages as cleaners because they contain sugars and other ingredients that will leave a stain.

Ammonia (ammonium hydroxide or aqueous ammonia). Ammonia has very irritating fumes, so never use it in a closed room. Diluted, it makes a good cleaner for glass or mirrors, and it dissolves grease. If mixed with detergent, it will remove wax. However, ammonia will damage asphalt tile, wood, and aluminum by pitting, staining, or eroding them. It will also change the color of some dyes.

Baking soda (sodium bicarbonate). Baking soda is an abrasive cleaner, but it is less likely to scratch surfaces than commercial products that contain silicon dioxide abrasives. It absorbs odors easily and is often recommended for odor removal.

Hydrogen peroxide. Hydrogen peroxide is used as a disinfectant because it helps kill germs. It is commonly used as a mild bleach for human hair, wool, or other delicate fibers.

Lemon juice (citric acid). Lemon juice is sometimes used as a mild bleach, but it can damage wool, cotton, linen, and rayon. It may be helpful in removing fresh, light, iron rust stains, but it may be ineffective on old or set stains.

Lye (sodium hydroxide). Lye is very toxic and can cause severe burns to the skin. It is a hazardous chemical, so use it with extreme caution.

Oxalic acid. Oxalic acid is often recommended for removing rust stains. Although it is effective for this purpose, it is not easily available in stores and is poisonous.

Salt. Sodium chloride, or salt, is sometimes recommended as an abrasive for cleaning, but it can scratch some metals. Salt does not "set" fabric dyes to prevent color bleeding in laundry.

Vinegar (10 percent acetic acid). Vinegar will cut grease film, but it will also pit some surfaces, such as marble. Vinegar may damage cotton, linen, and rayon or change the tint of some colored fabrics. Also, it does not "set" fabric dyes to prevent color bleeding in laundry.

Prepared with Joseph Ponessa, housing and energy specialist, Rutgers Cooperative Extension; and Brenda Cude, consumer economics specialist, UI Cooperative Extension Service.

Conserving energy and water

The environment and economics *can* go together. After all, many things that keep warm air from escaping from your house in winter also keep money from escaping from your checking account. Energy conservation can also mean a cleaner environment, as pollution is almost always generated when energy is processed for our use.

This section provides practical ideas on how to reduce your energy and water consumption. Your monthly power and water bills should help illustrate how well these ideas work for you.

51 Design new homes for energy efficiency

Homeowner savings

More than one-fifth of the energy produced in the world is used to heat and cool houses and other structures. Although home energy usage has not increased as rapidly in recent years, the cost of fuels such as oil, natural gas, and electricity continues to rise because the world's supply is limited.

Managing our rate of energy use will extend our energy supplies and slow the rate at which prices rise. And by incorporating energy-efficient building designs, you can save further without making major changes in your lifestyle or house.

Planning for savings and comfort

Key energy-saving points in designing a new home:

Building shape and size. A square building has less wall area per square foot than does a rectangular building of the same size; therefore, a square building has less heat loss per square foot. Also, a larger building usually has less heat loss per square foot.

Building orientation. Placing a rectangular house with the long axis east to west will mean less solar heat gain in the summer and more solar heat gain in the winter.

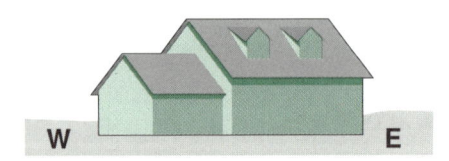

Roof overhang. A properly designed roof overhang on the south wall will protect the wall from summer sun without blocking winter sun.

Vestibule. A vestibule with an inner door reduces the flow of air in and out of the living area. This works especially well in homes with heavy traffic due to children and pets.

Window areas. For a given area, windows and doors lose more heat than walls do. A double-glazed window can lose seven times as much heat as an insulated wall.

East- and west-facing glass. In the summer, the amount of unwanted solar heat gained through unshaded east and west windows is twice as great as through equal-sized south windows.

Sill sealer. Placing a compressible filler on the foundation before the sill plate is fastened will reduce air infiltration. On older houses, caulk the joint between the sill and the foundation.

Roof surfaces. A light-colored roof reflects more solar heat than a dark-colored roof. This is not as important if the attic is properly insulated and ventilated. A flat roof can be sprayed with a reflective paint or covered with white stone chips.

Attic ventilation. Install continuous vents in the roof overhang and on the roof ridge to remove solar heat and eliminate the need for power venting.

Room over garage. The floor of a room over a garage should have 6 to 9 inches of insulation and a vapor retarder. The ceiling should have fire protection, such as Type X gypsum drywall.

Porches. A wide porch serves as an outside living space and shades the walls and windows against solar heat gain.

Crawl spaces. Cover the entire soil surface in the crawl space with a polyethylene vapor retarder. This prevents moisture from evaporating in the crawl space and moving as a vapor into the house.

Inside chimney. A chimney on an inside wall loses less heat to the outdoors and provides a better draft than a chimney on an outside wall.

Windbreak planting. Winter winds increase heat loss from buildings. A tall fence or a dense growth of evergreens will reduce the wind impact.

Shade planting. Tall deciduous trees will reduce solar heating of walls, windows, and roofs in the summer.

CHECK IT OUT

For more details on using plants to save energy: page 279.

Prepared with Mike McCulley, professor of architecture, UI School of Architecture—Building Research Council.

52 Save energy in heating and cooling

Saving energy in the winter

- Set thermostats wisely. A nighttime setback of 10 degrees may save as much as 7 percent in fuel. However, don't set back the thermostat in below-zero weather because reheating in the morning takes too long and uses a lot of energy. Also, when you're on vacation, don't set the thermostat below 50°F. Doing so could cause your pipes to freeze.
- Close unoccupied rooms and partially close registers to keep these rooms cooler—at about 50°F.
- Close the damper when the fireplace is not in use.
- Install glass doors on fireplaces to prevent heat from escaping up the flue. An outside opening for air intake to the fireplace will also reduce the amount of house air flowing up the flue when you use your fireplace.
- Caulk and weatherstrip the house. Sealing air leaks will not make your house too "tight." There are enough pathways for the entry of fresh air in most homes to eliminate the need to open doors and windows. Also, vent fans in kitchens and baths can provide the needed ventilation.

> **✓ CHECK IT OUT**
>
> For details on how and where to use caulking and weatherstripping: page 258.

Saving energy in the summer

- Make sure the air conditioner's compressor-condenser unit is shaded by a building, fence, or plantings. When shading the compressor-condenser, however, allow enough space for air movement around the unit.
- Close air supply registers in the basement so that you're not cooling the basement.
- Close doors on stairwells and room doors on the upper levels to control the downward flow of cooled air.
- If possible, hang clothes outside instead of using a clothes dryer. (In addition to saving energy, this may reduce wrinkling.)

- When practical, delay running dishwashers and other heat-producing appliances until late evening when power loads are less and air conditioners are more efficient.
- Shade windows to reduce the heat gain from solar energy. Shading methods rank in this order of effectiveness: (1) tree shade; (2) external shutters; (3) interior louvered screens; (4) interior window treatment, such as shades and drapes; and (5) tinted glass.
- Leave storm windows closed except on windows needed for ventilation.

High-efficiency heating systems

Today's high-efficiency furnaces are stingy on fuel use but will still keep you comfortable. Gas and oil furnaces manufactured since 1992 must be at least 78-percent efficient to meet federal regulations, and some units operate in the 90-percent efficiency range. The efficiency gauge, or Annual Fuel Utilization Efficiency (AFUE) rating, tells how much fuel is turned into heat and how much is wasted.

As a general rule, it can pay to replace furnaces made before 1965 because their efficiency is usually 55 percent or lower. If you have a working furnace that is 65-percent efficient or more, it is not as easy to justify replacement based on efficiency alone. Also, note that as efficiency increases, so does the price and complexity of servicing.

High-efficiency cooling systems

Both window and central air conditioners manufactured since 1992 are more efficient than older units. Central air-conditioning units are required to have a minimum Seasonal Energy Efficiency Ratio (SEER) of 10. What's more, some new air-conditioning units have a SEER that exceeds 15, and geothermal-based systems can reach as high as 20. But as efficiency increases, so does price.

Even if you have an older unit that is still operating, you may be justified in replacing it because you could save 30 to 40 percent in operating costs. However, replacement units should be properly sized because overcapacity units may not properly control interior humidity.

Heat recovery ventilators

Ventilation is necessary for health, but it will increase your heating and cooling bills. For the most energy-efficient way to ventilate your home, consider a heat recovery ventilator (HRV), also known as an air-to-air heat exchanger.

With an HRV, there are two air flows—one into the house and one out of the house—but the streams never mix. During the summer, as the two streams pass through the exchanger, the cool outgoing air extracts heat from the incoming air. As a result, the HRV cools the incoming air, reducing the load on the air conditioner.

During the winter, the cool air entering the house draws heat from the warm outgoing air. By heating the incoming air, the HRV reduces the load on the furnace.

HRVs are popular in Europe and Japan, but they are still relatively new for home use in the United States. If you have trouble finding one, or if you have second thoughts about using an HRV, don't give up on ventilation. You should still ventilate with a traditional exhaust ventilator.

Also, keep in mind that HRVs call for conscientious maintenance.

Speaking of maintenance...

When serviced annually, your air-conditioning and heating systems will operate more efficiently, and you may delay the need for replacement. For warm-air systems, have a competent serviceperson make these adjustments each year:

- Check and clean burners.
- Oil the fan and motor bearings.
- Inspect the fan belt (if there is one), and replace or clean air filters.
- Vacuum return-air grilles as necessary.
- Adjust thermostat controls for proper air circulation during cold weather. Set the furnace so the blower will kick in when air inside the furnace reaches 110°F. Set the blower to turn off when the air inside the furnace cools to 85°F.

Prepared with Mike McCulley, professor of architecture, UI School of Architecture—Building Research Council.

Minneapolis

Cutting the cost of keeping out the cold

The unofficial state slogan in Minnesota could very well be, "Cold enough for you?" So it is only fitting that builders in Minnesota have been at the forefront of demonstrating ways to cut heating costs by one-third to one-half.

The University of Minnesota's Cold Climate Housing Project, along with builders and a variety of organizations, have constructed several model homes that show you can tighten a building for energy efficiency without keeping pollutants bottled up inside the home. In fact, you can *improve* indoor air quality while saving energy.

One of these model homes, constructed in 1990 in a suburb of Minneapolis–St. Paul, has drawn considerable attention, says Pat Huelman, coordinator of the Cold Climate Housing Project.

"The techniques used in the home are not rocket science features, but in this market they were fairly new at the time," Huelman says. "Today, we are beginning to see most of these features showing up in new construction."

Huelman highlights some of the most significant features, beginning with the foundation and working his way up:

Radon protection
A continuous polyethylene sheet under the basement floor reduces the risk of radon seeping into the basement from the soil.

Foundation insulation and waterproofing
The foundation is insulated with semirigid fiberglass panels. Because the fiberglass fibers run vertically, moisture runs down them, drains into tiles, and is then taken to a sealed pump.

Walls

The walls are insulated with high-density fiberglass batts. In addition, the advanced framing system in the walls minimizes the amount of framing materials and reduces the number of gaps and breaks in the insulation—a key to improving the effectiveness of insulation.

Another key is to keep insulation dry. Builders did this with a vapor barrier of durable cross-laminated polyethylene. They also eliminated a lot of the gaps around electrical outlets using a "poly pan"—a plastic box in which the electrical box fits. The vapor barrier attaches directly to the poly pan.

The exterior of the walls is protected with a house wrap made of spun polyethylene to keep out both water and wind.

Windows

The windows are "quad pane." This means each window consists of two panes of glass with two "low-emissivity" films between them. Low-emissivity films reduce the movement of heat into and out of the house.

The gaps between the two panes of glass and the two films are filled with an argon/krypton gas mixture, which reduces the movement of heat through the window.

Air-to-air heat exchanger, or "heat recovery ventilation"

Air exhausted from the bathroom, kitchen, and laundry room is linked to the heat exchanger. As this indoor air is vented outside, it heats or cools the fresh air coming into the home—reducing the work load of the furnace or air conditioner.

High-efficiency furnace

When a furnace burns more efficiently, the combustion process produces more water vapor and less waste heat. In tightly sealed homes, this can result in poorer draft up the chimney. However, high-efficiency furnaces, such as the one installed in this Minnesota home, are designed to prevent these problems. They can handle the added condensation, and they power-vent the byproducts of combustion directly outside.

In contrast, mid-efficiency furnaces (82- to 88-percent efficient) are often susceptible to draft problems, and early models couldn't handle the added moisture (which can corrode the furnace).

Sealed combustion systems

The fresh air needed for the combustion process in most furnaces is normally provided by air inside the house. But with a sealed combustion system, this fresh air comes from outdoors.

One benefit is no risk of "back drafting." Back drafting is when one combustion or exhaust appliance pulls pollutants from another combustion appliance's flue, drawing them into the house. (For more details, see page 226.)

High-efficiency air conditioner

The air conditioner is also a high-efficiency unit. The forced-air heating and cooling system includes an air filtration system that removes dust and mold particles.

Sealed gas fireplace

The fireplace features a gas log system with a sealed glass door and direct vent system that reduces the potential for combustion contaminants entering the house. It also reduces the risk of back drafting.

Sealed air ducts

All seams on the ducts are sealed to prevent leaks. This also lowers the "negative pressure" in a basement, which reduces the potential for radon to enter the basement.

Many of these ideas can be adapted to an existing home, Huelman says, although some of them would require major renovations.

But can the savings in energy with these systems make up for their cost?

According to Huelman, this system cost approximately $4,000 for all components, and the home was projected to heat for less than $200 for the entire year.

"The owner of this home could expect a payback in seven to ten years," he estimates. "That's a reasonable payback."

53 Insulate and seal your home

Types of insulation

The insulation that you apply to walls, ceilings, and foundations wraps a house, just as a warm coat wraps your body. In the process, insulation saves fuel, saves money, and makes your home more comfortable.

All insulating material is rated according to "R-values," which indicate a material's ability to prevent heat from flowing into or out of the house. The higher the R-value, the more effective the insulation.

There are three basic types of home-insulating materials:

Batt insulation. Batt insulation is sold in rolls or blankets of various thicknesses, and most of it is made of fiberglass. It is used most often in building cavities (walls and ceilings) and has R-values that range from 3.1 to 4.3 for every inch of thickness.

Batt insulation is moisture-resistant, but not very resistant to vapor. Therefore, apply a polyethylene vapor barrier on the warm side of batt insulations. Some batt insulation comes with a built-in vapor barrier.

Loose-fill insulation. Loose-fill insulation can be made of fiberglass, rock wool, or cellulose treated with a fire retardant. (Cellulose is made from wood products such as newspaper.)

Loose-fill insulation has an R-value that ranges from 2.2 to 3.7 per inch, and it is often blown into attics and walls. It is resistant to fire (assuming cellulose has been treated with a fire retardant), but moisture can reduce its effectiveness.

Rigid foam boards. Rigid foam boards are made of different types of plastics and are usually sold in 4-foot by 8-foot or 2-foot by 8-foot panels. The R-values for foam boards range from 3.8 to 8 per inch.

Rigid foam is often used in new construction to cover the outside of walls. It is also a good choice for insulating finished basement walls and the outside of a foundation. However, fire is a concern. Some plastics are highly flammable and produce toxic smoke when burned. For this reason, building codes require a barrier, such as drywall, to protect this insulation from heat and flame.

Weatherstripping and caulking

Plugging holes and cracks with weatherstripping and caulking are easy ways to keep drafts out and lock heated or cooled air inside your house. Even a small crack can let in a large amount of air. For instance, a 36-inch by 80-inch door with a $\frac{1}{32}$-inch crack (thinner than a dime) around the edges lets out as much heat as a $7\frac{1}{4}$-square-inch opening.

Insulation values recommended for key areas in the house

Location	Recommended R-value for insulation
Ceilings over heated areas	R-30 or greater
Exterior walls of a heated area	R-19 or greater
Floors over vented crawl space[a]	R-19
Basement walls	R-11
Crawl-space walls[a]	R-10 with vapor retarder material on the ground floor
The "rim" or "band joists"[b]	R-19

[a]If you insulate the floor over a vented crawl space, you do not need to insulate crawl space walls—and vice versa.
[b]This is the space above the foundation wall, in either a crawl space or basement, where the joists sit on top of the foundation. A single "band joist" covers the ends of the joists.

Consider the following options.

Weatherstripping options. Here are different types of weatherstripping:
- Spring metal is very durable, but somewhat difficult to install.
- Adhesive-backed foam is not durable, but it is easy to install. Peel away the backing and stick it to clean, dry surfaces.
- Felt may deteriorate, so do not use it outside. It is easy to install.
- Tubular gaskets and vinyl tubing are durable and easy to install.
- Foam-edged wood is more durable than foam strip alone. But it cannot be used on windows.

The accompanying illustrations show where to attach these different types of weatherstripping to windows and doors.

Caulking options. Use caulking compounds to fill and cover cracks and holes in these locations:

Inside...
- Around door and window frames
- Along baseboards
- Around wall outlets
- Around ceiling fixtures and all other openings to the outside

Outside...
- Between the window frame and siding
- Between the windowsill and siding
- Where the storm window meets the window frame (except at the windowsill)
- Between the door frame and siding

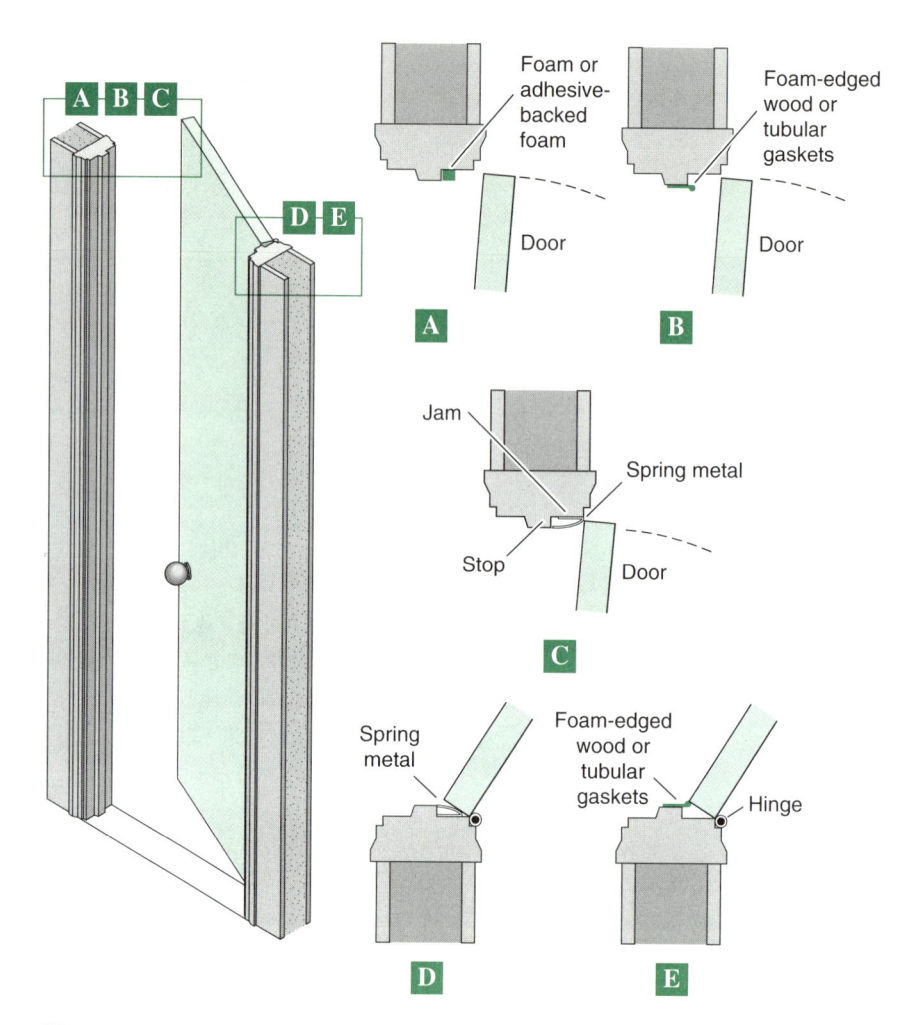

Weatherstripping on doors

A	B	C
Between jamb and edge of sash	Top of upper sash	At meeting rail between sashes

Spring metal strips on windows

You can weatherstrip double-hung windows with spring metal strips, vinyl weatherstripping, and adhesive-backed foam. Pictured above are some of the key locations to place spring metal strips. Another good spot for spring metal weatherstripping is along the bottom of the lower sash.

- At corners formed by siding
- Where the wood structure meets the foundation
- Between the porch and the main body of the house
- Where the overhang meets masonry walls
- Where the chimney or masonry meets the siding
- Where the chimney meets the roof
- Around outside water faucets, dryer vents, and other openings
- Where telephone service enters the house
- Where pipes and wires penetrate the ceiling below an unheated attic

Do not caulk the holes commonly found in vertical mortar joints along the bottom of brick walls. These "weep holes" are there to drain moisture that penetrates the wall.

The following are the most common caulking compounds:

- Oil or resin bonds to most surfaces and lasts up to seven years. It has fair to good adhesion and should be painted.
- Latex, butyl, and polyvinyl last up to ten years and have good to excellent adhesion. When used indoors, painting the caulk is optional; when used outdoors, coat the caulking with latex paint.
- Silicons, polysulfides, and polyurethanes last more than twenty years and have excellent adhesion. Silicon is good to use around a bathtub

Sealing air leaks by caulking around windows usually does not make a house too "tight." In most homes, there are enough paths for fresh air to enter without having to open windows and doors—except when you need to reduce odors or humidity. However, it's a good idea to test for carbon monoxide and radon before and after weatherization work.

because it has high moisture resistance. These are the most expensive caulking products, and some cannot be painted.

Insulating windows

When you are shopping for windows, you are likely to find their energy efficiency rated by "U-values," rather than R-values. While R-values measure the ability of insulation material to resist the flow of heat, U-values measure the ability of a window to *conduct heat* to the outside. When comparing insulation, you want to look for high R-values; but when you're looking at windows, you want to find glass with the *lowest* U-value. The lower the U-value, the less a window will conduct inside heat to the outside.

The poorest insulator in your house is single-pane glass. Adding layers of glass can help reduce heat loss, because as more panes are added, the U-value decreases. Double glazing (two layers of glass) cuts heat loss almost in half, and triple glazing cuts heat loss by two-thirds (but it may also decrease light entering the house by 10 percent).

You can double-glaze a window in various ways:

- Placing a storm window over a single-pane window
- Replacing a single-pane window with an insulated window that consists of two layers of glass separated by space

U-values for various types of windows

Glazing	U-value
Single glazing	0.94
Double glazing	
Single-pane plus storm window	0.48
Double-pane (sealed glass, ½" air space)	0.48
Double-pane (wavelength-selective sealed film or glass, ½" argon gas-filled)	0.39
Triple glazing	
Double-pane (sealed glass, ½" air space) plus storm window	0.38
Triple-pane (sealed glass, ½" air space)	0.38
Triple-pane (sealed glass, wavelength-selective film or glass, ½" argon gas–filled)	0.27

NOTE: The lower the U-value, the better the insulation provided by the window.
U-values for actual windows may vary due to glass thickness, frame type, and coatings.
SOURCE: 1993 ASHRAE Fundamentals.

Single glazing
Single-pane window

Double glazing
Double-pane window

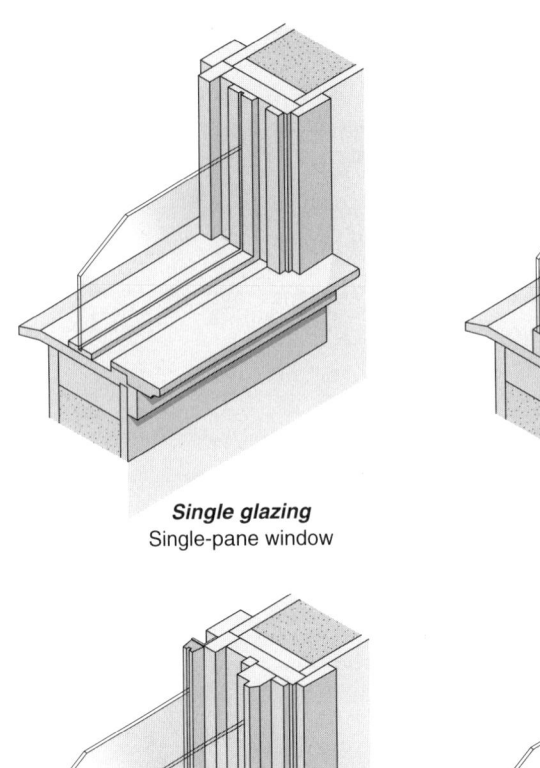

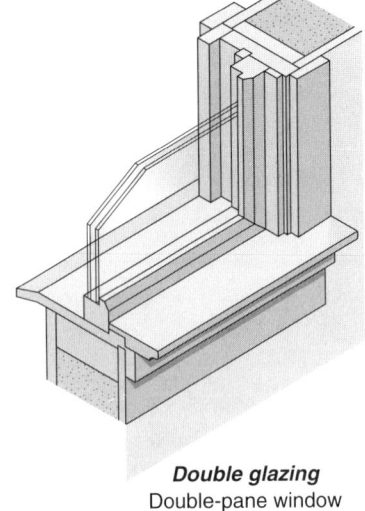

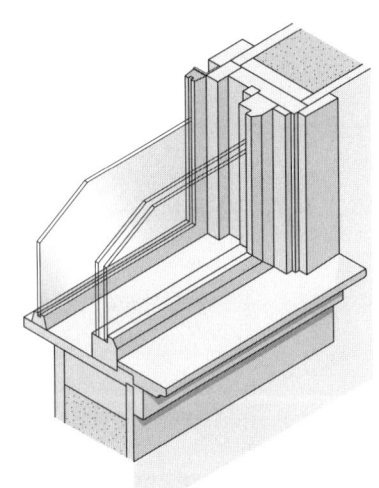

Double glazing
Single-pane window
plus storm window

Triple glazing
Double-pane window
plus storm window

Double glazing (two layers of glass) cuts heat loss almost in half, and triple glazing cuts heat loss by two-thirds.

- Attaching a separate glass and sash on the inside of the single-pane window
- Attaching a plastic film or sheet to the window trim

What about indoor air pollution?

By tightening up a house with insulation, weatherstripping, and caulking, you may reduce the number of "air changes per hour" in a home. In turn, by reducing the number of air changes in the home, you can sometimes increase the level of indoor pollutants, including carbon monoxide from combustion appliances and fireplaces. Therefore, whenever you tighten up a house with major weatherization improvements, test for carbon monoxide and other pollutants (such as radon) before and after the weatherization work.

CHECK IT OUT

For more information on weatherization and indoor air pollution: page 226.

Prepared in cooperation with the UI School of Architecture—Building Research Council.

54 Choose energy-efficient appliances

Efficiency pays

Over the life span of most appliances, the cost to operate them is far more than the initial purchase price. Therefore, purchasing an energy-efficient appliance is almost always the wisest choice, even if it costs more to purchase than an inefficient one.

The top ten energy users in the typical U.S. home (not including the central heating unit) are:

1. Water heaters
2. Refrigerators
3. Freezers
4. Air conditioners
5. Ranges
6. Clothes washers
7. Clothes dryers
8. Dishwashers
9. Portable space heaters
10. Lights

The higher the appliance is on the list, the more sense (and cents) it makes to look for the most energy-efficient model possible.

'EnergyGuides'

When shopping for an appliance, you can determine which model is the most energy-efficient by looking at the "EnergyGuide" labels. EnergyGuide labels were revised in 1995, so you may see either the old or new labels. Under the new system, EnergyGuide labels fall into two basic categories—*energy use* labels or *energy efficiency* labels (see illustrations).

Energy use labels. Energy use labels appear on refrigerators, freezers, water heaters, dishwashers, and clothes washers. The amount of energy used per year will be listed in a rectangle in the center of the label. If the product is an electrical appliance, the label will list the number of "kilowatt hours per year" that the model uses; if it is a gas appliance, it will list the number of "therms per year" that the model uses. The *lower* the number, the more energy-efficient the appliance.

The energy use label will also show how these levels of energy consumption compare to other models. It will show where the model falls along a scale that ranges from the most energy-efficient to the least energy-efficient. Finally, the energy use label will show how much it costs to operate the appliance for a year, based on the national average cost for electricity and gas.

Energy efficiency rating labels. Climate control appliances, such as room and central air conditioners, furnaces, and heat pumps, are labeled with an energy efficiency rating. If the appliance is an air conditioner or heat pump, the label will list the Seasonal Energy Efficiency Ratio (SEER); if the appliance is a furnace or boiler, the label will list the Annual Fuel Utilization Efficiency (AFUE). With both ratings, the *higher* the number, the more efficient the appliance.

Energy efficiency rating labels also show a scale ranging from the least efficient to the most efficient model available; and they show where the appliance you're considering falls within that range. Ask the seller or installer of the appliance for a fact sheet or directory for more information about the efficiency and operating cost of the appliance.

Annual and life-cycle costs

Energy use labels provide estimated costs to operate the appliance for a year. But those costs are based on a *national* average for energy rates. For a more accurate estimate of how much the appliance will cost to operate annually, check your utility bill to find out either the kilowatt-per-hour rate (for electricity) or the therm rate (for gas). Then multiply those rates by the kilowatts or therms per year that the appliance will use.

Example: Assume your local utility charges 12 cents per kilowatt hour, and the EnergyGuide for the refrigerator you're considering says it uses 776 kilowatt hours per year. Multiply the two numbers and you find that the appliance will cost $93 to operate annually.

You can also compute how expensive it will be to operate an appliance *over its entire life,* allowing you to judge how much one model will save over another. To compute the lifetime costs, multiply the per-year operating cost times the average life span of the appliance, and add that number to the original price. According to industry officials, the average life span of major appliances is estimated at:

Refrigerator	20 years
Freezer	20 years
Clothes dryer	18 years
Range/oven	18 years
Room air conditioner	15 years
Clothes washer	13 years
Water heater	13 years
Central air conditioner	12 years
Dishwasher	12 years

Based on standard U.S. Government tests

ENERGYGUIDE

Refrigerator-Freezer
With Automatic Defrost
With Side-Mounted Freezer
Without Through-the-Door-Ice Service

XYZ Corporation
Model ABC-W
Capacity: 23 Cubic Feet

Compare the Energy Use of this Refrigerator with Others Before You Buy.

This Model Uses
776 kWh/year

Energy use (kWh/year) range of all similar models

Uses Least
Energy
776

Uses Most
Energy
1476

kWh/year (kilowatt-hours per year) is a measure of energy (electricity) use. Your utility company uses it to compute your bill. Only models with 22.5 to 24.4 cubic feet and the above features are used in this scale.

Refrigerators using more energy cost more to operate. This model's estimated yearly operating cost is:

$64

Based on a 1992 U.S. Government national average cost of 8.25¢ per kWh for electricity. Your actual operating cost will vary depending on your local utility rates and your use of the product.

Important: Removal of this label before consumer purchase is a violation of Federal law (42 U.S.C. 6302).

An energy use label shows how much energy (in this case, electricity) that an appliance uses annually. It also provides an estimate of an appliance's annual operating cost.

ENERGYGUIDE

Based on standard U.S. Government tests

Furnace—Natural Gas

XYZ Corporation
Model 2345X

Compare the Energy Efficiency of this Furnace with Others Before You Buy.

This Model's Efficiency
80.7 AFUE

Energy efficiency range of all similar models

Least Efficient	Most Efficient
78.0	97.0

The AFUE, Annual Fuel Utilization Efficiency, is the measure of energy efficiency for furnaces and boilers. Only furnaces fueled by natural gas are used in this scale.

Natural gas furnaces that have higher AFUEs are more energy efficient.

Federal law requires the seller or installer of this appliance to make available a fact sheet or directory giving further information about the efficiency and operating cost of this equipment. Ask for this information.

Important: Removal of this label before consumer purchase is a violation of Federal law (42 U.S.C. 6302).

An energy efficiency rating label rates appliances on a scale that ranges from the least efficient to the most efficient model available.

Example: Let's assume you had narrowed your choice of a refrigerator down to two models—one with a purchase cost of $800 and an annual energy cost of $120 and another, more energy-efficient model with a purchase cost of $1,000 and an annual energy cost of $93. Here is what those refrigerators would cost if you included estimated operation costs over their entire life:

Refrigerator #1: $800 + ($120 annual energy costs x 20 years)=$3,200

Refrigerator #2: $1,000 + ($93 annual energy costs x 20 years)=$2,860

Thus, the energy-efficient model with the higher purchase price will cost about $340 less to purchase and operate in the long run.

For a specific list of the most energy-efficient models of a particular appliance, you may want to consult such publications as *Consumer Reports*, especially its annual buyers guides, and the *Consumer Guide to Home Energy Savings* by the American Council for an Energy-Efficient Economy. For information on the *Consumer Guide to Home Energy Savings*, write to 2140 Shattuck Ave., Suite 202, Berkeley, CA 94704.

Maintaining your appliances

Increased energy efficiency also comes with the proper maintenance of appliances. Follow these guidelines.

All appliances

- Read and follow the instruction booklet to find out how to properly maintain the appliance.
- Inspect the electric cords and hookups to make sure they are in good condition.
- Clean the appliance regularly.

Dryers

- Proper venting is critical for good results. Make sure there is no obstruction in the exhaust duct or hinged hood cover that restricts air flow. Clean out the duct periodically.
- Clean the lint filter after every load so the dryer's air flow is not obstructed.

Refrigerators

- Make sure the refrigerator is standing level.
- Make sure the refrigerator has sufficient ventilation and air circulation.
- Clean the condenser fins and coils at least every three months.

> ✓ **CHECK IT OUT**
> For information on the maintenance of air conditioners and furnaces: page 254.

The next generation: The eco-fridge

In 1994, a coalition of 24 utility companies awarded $30 million to the company that designed the most energy-efficient refrigerator. The winning company came up with a refrigerator that cut energy use by 30 percent when compared to minimum government requirements; and it has been collecting its prize money in bits and pieces—$100 for every energy-saving appliance that reaches the stores.

Over the years, appliances have become increasingly energy-efficient, but this contest symbolized a new level of interest. Many companies are scrambling to outdo the other in energy efficiency, so keep your eyes open for these new eco-appliances. Then compare prices and energy savings.

Prepared with Mary Ann Fugate, consumer and family economics educator, and Brenda Cude, consumer economics specialist, UI Cooperative Extension Service.

55 Save money and energy with compact fluorescent light bulbs

A 200-percent return

If you had the opportunity to make an investment with your money that offered a return of at least 200 percent, you would probably be enticed to explore the possibility. Well, such an opportunity already exists, but not where you might expect it. The opportunity is in the light bulbs you use to light your home, office, workshop, and other areas.

Compact fluorescent light bulbs not only save you money but also help reduce the consumption of hard-to-replenish natural resources, such as coal, oil, and natural gas. In turn, they also help reduce the emission of gases that result when utilities burn these fossil fuels to power electricity-generating facilities. Some scientists believe that the gases, such as carbon dioxide, sulfur dioxide, and nitrogen oxides, might be increasing the Earth's greenhouse effect, possibly contributing to or magnifying changes in global climate.

At first glance, switching to compact fluorescent light bulbs might not look economical, even though it would be environmentally responsible. When you compare the $22 price tag for one compact fluorescent bulb (including the ballast) with the 50-cent price for one ordinary incandescent bulb, the benefits may seem hard to imagine. But the payoff comes in the longevity of compact fluorescent light bulbs and in their stinginess with electricity.

As you can see in the accompanying chart, a 27-watt compact fluorescent light bulb is a little more expensive over the first 1,000 hours of operation when compared to its equivalent incandescent bulb. But by the time it has burned for 12,000 hours (the life expectancy of one compact fluorescent light bulb), you can potentially save about $54—a 245-percent return on your investment. After 65,000 hours (the life expectancy of a ballast plus five to six replacement bulbs), you can potentially save $345.

If you multiply these amounts by the number of light fixtures in your house, the potential savings can really add up.

Dollar savings with compact fluorescent light bulbs
27-watt compact fluorescent vs. 100-watt incandescent

Type of bulb	Kilowatts needed	Cost of electricity[a]	Cost of light bulbs[b]	Total cost	Savings with compact fluorescent bulbs
1,000 hours of operation					
Incandescent (100W)	100	$8.00	$.50	$8.50	None
Compact fluorescent (27W)	27	2.16	22.00	24.16	
12,000 hours of operation					
Incandescent (100W)	1,200	$96.00	$6.00	$102.00	$54.08
Compact fluorescent (27W)	324	25.92	22.00	47.92	
65,000 hours of operation					
Incandescent (100W)	6,500	$520.00	$32.50	$552.50	$345.10
Compact fluorescent (27W)	1,755	140.40	67.00	207.40	

NOTE: A 27-watt compact fluorescent light bulb is the equivalent of a 100-watt incandescent light bulb. Also, 1,000 hours is the typical life span of an incandescent bulb burned for about 5 hours per day; 12,000 hours is the typical life span of a compact fluorescent light bulb burned for about 5 hours per day.
[a] Calculated at 8 cents per kilowatt hour.
[b] Calculated at 50 cents per 100-watt incandescent bulb. Estimates for the compact fluorescent light bulbs include the initial cost of the ballast, plus the cost of replacement bulbs after 12,000 hours. The figure used is an average; some stores carry 27-watt compact fluorescent light bulbs for as little as $17.

Design

Compact fluorescent light bulbs come in two basic designs:
- Integral, or one-piece. The light bulb and the ballast are fused into one piece. The ballast is the electricity-regulating device that is part of all fluorescent lights.
- Modular, or two-piece. The ballast and light bulb are separate. Because ballasts usually last five to six times as long as the light bulbs, even greater savings are possible with two-piece compact fluorescent lights; you will not have to replace the ballast every time you replace the bulb.

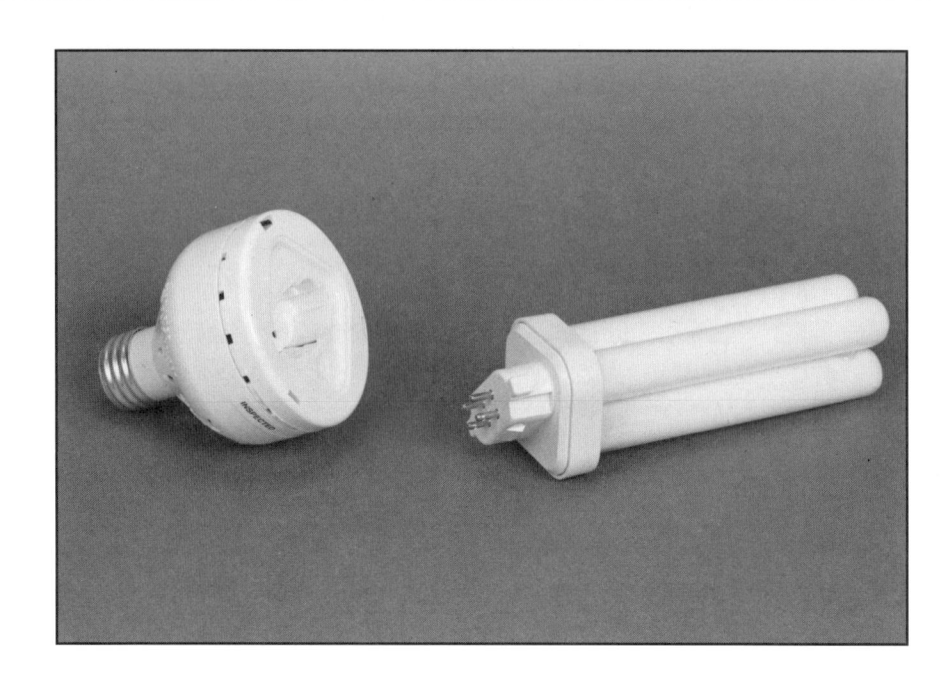

Here are two views of a compact fluorescent bulb with a detachable ballast. With this kind of bulb, you don't have to replace the ballast every time you replace the bulb.

Ballasts

There are two basic choices of ballasts for compact fluorescent light bulbs: electric or magnetic. Ballasts used to be only available in the magnetic design. These are the kind used in "old-fashioned" fluorescent lamps, which flickered before coming fully on, took some time to achieve full brightness, and were often accompanied by an annoying buzz. New electric ballasts are silent, turn the lights on instantly, and are virtually flicker-free.

Another drawback of the magnetic ballasts is that some contain small amounts of radioactive materials. Although these materials are well contained, they still represent a waste disposal problem and a possible health hazard.

Style

The two most common styles of compact fluorescent bulbs are circular and vertical. Depending on the lamp and lamp shade, each style has advantages over the other. For example, some vertical compact fluorescent light bulbs will not fit under the mounting bracket that holds the shade on certain lamps. Although this problem has been greatly reduced with the advent of ultracompact vertical fluorescent bulbs, circular bulbs often fit better than vertical bulbs in lamps originally designed for incandescent bulbs.

Some compact fluorescents come in more ordinary-looking shapes—ideal for exposed bulbs in places such as basements and garages.

Some companies produce "ultracompact" vertical fluorescents, which are easier to fit in lamps. The one drawback of an ultracompact is that the ballast is not detachable.

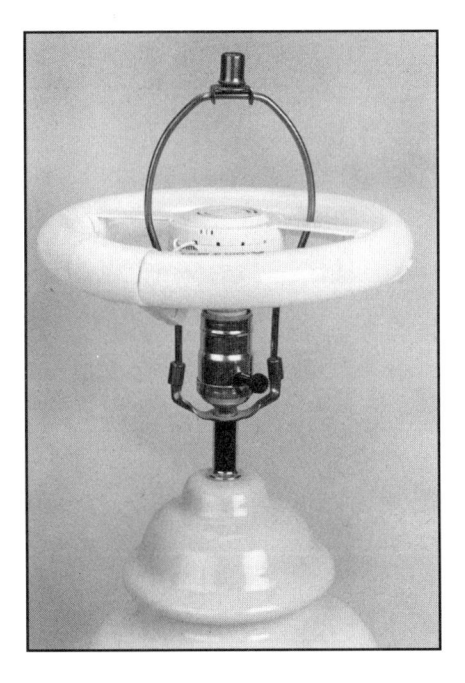

To fit a circular compact fluorescent bulb in your lamp, attach the shade frame on the inside of the circle.

Color rendition index

The color rendition index (CRI) refers to how true colors look under a particular light source. Although some people do not like the "color" of light produced by fluorescent lights, many of the newest compact fluorescent light bulbs actually produce a light that is a close match to natural sunlight; they have CRIs as high as 80 to 90 out of a possible 100. However, some are also as low as 60. Natural sunlight is 100 and incandescent bulbs have CRIs of about 99.

CRI values are not always printed on the box that the bulb comes in, but they are available from the manufacturer.

How significant is the impact on the environment?

For every lamp in which you replace a 100-watt incandescent bulb with a 27-watt compact fluorescent light bulb, you will contribute to a decrease in carbon dioxide, sulfur dioxide, and nitrogen oxides emissions. The accompanying chart shows just how much you can reduce the emission of these gases.

Emission reductions with compact fluorescent light bulbs
27-watt compact fluorescent vs. 100-watt incandescent

	Kilowatts of electricity avoided	Carbon dioxide emissions avoided	Sulfur dioxide emissions avoided	Nitrogen oxides emissions avoided
1,000 hours of operation	73	109 lb.	1 lb.	0.5 lb.
12,000 hours of operation	876	1,314 lb.	11 lb.	5 lb.
65,000 hours of operation	4,745	7,118 lb.	60 lb.	26 lb.

NOTE: A 27-watt compact fluorescent light bulb is the equivalent of a 100-watt incandescent light bulb. Also, 1,000 hours is the typical life span of an incandescent bulb burned for about 5 hours per day; 12,000 hours is the typical life span of a compact fluorescent light bulb burned for about 5 hours per day.

It is mind-boggling to consider how big these numbers would be if lamps all over America were equipped with compact fluorescent light bulbs. In fact, the U.S. Environmental Protection Agency (EPA) is trying to accomplish just that with its Green Lights Program. For more information, call the Green Lights Hotline at (202)775-6650.

Energy-efficient compact fluorescent light bulbs are not yet available in retail stores in all areas. However, you can obtain them by mail, or you can get information on them from sources such as:

Feit Electric Company	(800)543-3348
GE Answer Center	(800)626-2000
Lights of America	(800)321-8100
Phillips	(800)631-1259
Osram	(800)544-4828
Real Goods	(800)762-7325
SK America, Inc.	(800)793-1212
Supreme Lighting Company	(800)221-1573

Prepared with Nicholas Smith-Sebasto, environmental education specialist, UI Cooperative Extension Service.

56 Use plants and other landscape tools to conserve energy

Shade, insulation, and windbreaks

What you plant and where you plant *outside* your home can have a major impact on the amount of energy you use to heat and cool the *inside* of your home. With an eye to energy conservation, evaluate the location of plants and other landscape features around your house. This will help you use landscape materials to do two things:

- Block summer sun and winter winds.
- Allow access to winter sun and summer breezes.

Keep in mind that trees and shrubs are not the only landscape tools that can provide shade and control wind. Earthen berms, or ridges, can channel summer breezes into certain areas or intercept the sun's rays. Earth can also be graded against a wall or structure to insulate a home. You can use slatted fences to control the wind, and you can use ground covers and vines to affect the temperature in and around a house.

Protecting the house from summer sun

Different walls of a house have different needs for shading.

The north and east walls. In the summer, the sun rises in the northeast and sets in the northwest. This wide swing of the sun permits a small amount of sunlight to strike the north wall in the early morning hours before bearing down on the east wall. The heating effect on the east wall is quite strong because of the low angle of the sun.

The sun's rays are almost perpendicular to the east wall, so there is maximum heat absorption by the wall surface. However, the air and the house are not heated significantly because of the cooling that took place overnight. For this reason, tree plantings are probably not necessary to shade the east wall unless a large area of glass is part of the wall.

The south wall. The south wall receives full radiation between 11 a.m. and noon. Depending upon the overhangs and window locations of your house, you may want to locate a tree to shade the south end of the east wall and the east end of the south wall. In general, however, the use of trees to shade the south wall is not effective because the shadow cast is minimal. A roof overhang does a much more efficient shading job.

At noon, the sun is at its highest angle with relation to the earth. The radiation from the sun hits the south wall at such a steep angle that the amount of heat absorbed by the wall is much less than might be expected (the steeper the angle, the less heat is absorbed by the house). In fact,

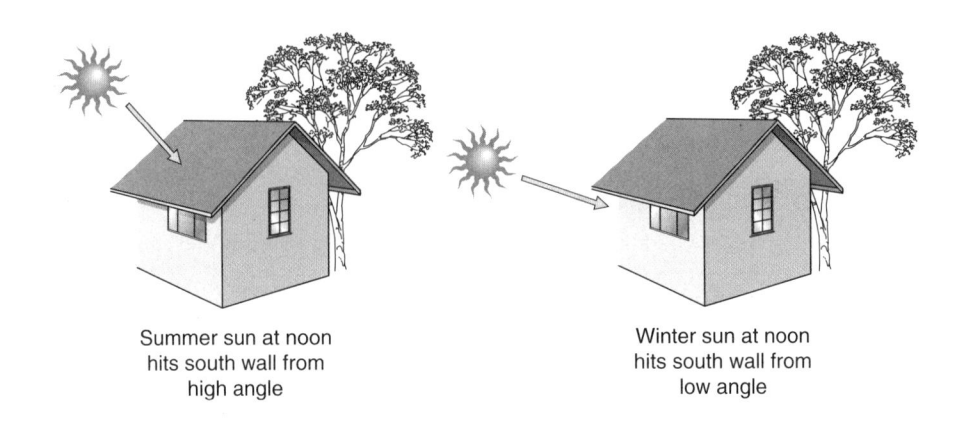

Summer sun at noon
hits south wall from
high angle

Winter sun at noon
hits south wall from
low angle

The steeper the angle of the sun, the less that heat is absorbed by a house.

summer midday sunshine delivers less heat against the south wall than the low-angle midday sunshine during winter.

The west wall. Between 1 and 2 p.m., the sun is starting to descend, and the west wall begins to receive radiation. A large tree can shade part of the roof area and the south end of the west wall. Because of the short shadow cast, plant such a tree as close as 15 feet from the house.

For the remainder of the afternoon, the west wall receives the same exposure that the east wall did earlier. The major difference is that the sun is at its maximum heating capacity, the air is hot, and the house has lost its night coolness. The sun is directly in the west by 4 p.m., the angle of the sun is low, and the direct radiation has powerful heating capabilities.

If there is space for only one tree to protect the west wall, place it up to 25 feet from the house on a line between the 3 p.m. and 5 p.m. sun positions. A distance of 25 feet from the house allows you to take advantage of the longer shadows cast by the low angle of the sun. A small formal or informal hedge or earth berm beneath the tree will help to intercept the low angle of the sun.

Another alternative is to use vines to shade the wall and absorb and reflect the heat. You can place vines on both frame and masonry houses.

What about the winter months?

In the winter, sunrise is in the southeast and sunset in the southwest. There is little sun on the east or west walls, and the south wall receives nearly all of the winter day's sunshine.

In the winter, you want to take full advantage of the sun's heat. Therefore, if you place a tree on the south wall as protection against the summer sun, make sure it is a deciduous tree, which loses its leaves in the fall. The sun can shine through the leafless branches during the winter.

Controlling the wind

You cannot stop the wind, but you can divert it, channel it, deflect it, and intercept it. Although trees and plants are commonly thought of as the only way to intercept wind, fences also work well. Fences have two advantages over other windbreaks: they require less ground space and they provide immediate protection.

If you use a fence as a windbreak, however, be sure it has slats. Solid fences do not stop the wind; they make the wind more vicious. You can further increase the effectiveness of a fence by adding fins or deflectors along the top edge.

Because most communities do not permit fences over 6 feet high, you will not have as large a zone of protection as with a higher evergreen windbreak. To determine how much protection you can get from a windbreak, multiply the windbreak's height by 20. This means that a 10-foot-high windbreak will give you a zone of protection of 200 feet (10 x 20 = 200 feet). There is at least some wind reduction in this zone of protection.

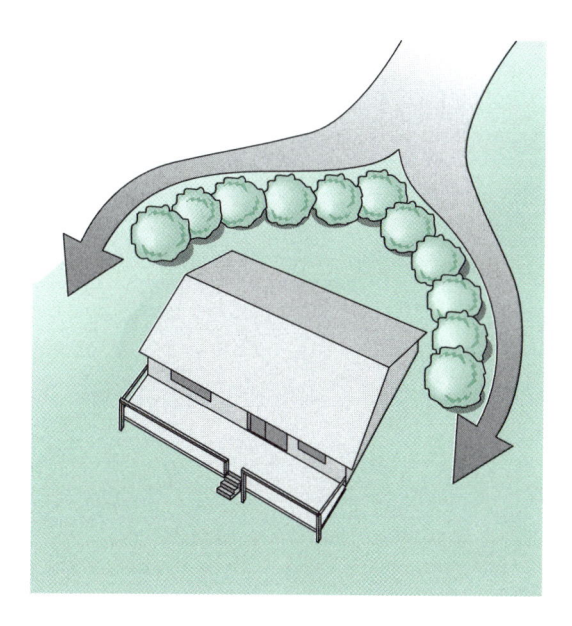

A windbreak can divert a cold air mass moving downhill.

Both deciduous and evergreen plants can be used for windbreaks, but if you are trying to protect against winter wind, use evergreens because deciduous plants lose their leaves during winter. In towns and cities, the Canadian hemlock is a highly desirable plant. It can be sheared both at the top and along its sides. The spacing between plants should be no more than 14 feet. If space permits, plant a maximum of three rows for the greatest protection.

Windows and entryways

Windows and entryways are prime sources for heat loss or gain. Evaluate their location according to the paths of winter and summer sun and prevailing winds and breezes. A baffle fence or courtyard development with small trees and shrubs will decrease heat gain or loss in doorways.

If a home has exterior walls that include a projection or recessed area, special attention is required. Projections and recessed areas are highly vulnerable to wind and air filtration, resulting in major heat gains when the house is being air-conditioned and heat losses when the house is being heated. Fences or plantings can reduce these problems.

Prepared with David Williams, horticulturist, UI Cooperative Extension Service.

Additional source: Designing an Energy-Efficient Landscape, *by William R. Nelson, Jr., University of Illinois Circular 1178.*

57 Conserve water in and around the home

Saving water, saving money

Conserving water means smaller water bills if you live in an area serviced by a municipal water system. This will be an increasingly important consideration in the future as consumers demand more protection from contaminants in their water supply—as monitoring increases, so do water costs.

If your home is connected to a septic system, water conservation also can help prevent water pollution. Overloading a septic system may cause nutrient and bacterial contamination of nearby lakes and streams, or it may contaminate your well water.

The first step in understanding how to conserve water in your home is to know where water is used most heavily. People typically use 45 to 50 gallons of water indoors each day and up to the same amount outdoors, depending on the season.

Studies show wide variations in different areas of the country and between urban and rural households. But according to the American Water Works Association, the average U.S. family uses water inside the home at the following rates:

Toilets	28 percent
Clothes washing	22 percent
Showers	21 percent
Faucets	12 percent
Baths	9 percent
Toilet leakage	5 percent
Dish washing	3 percent

Conserving water in the bathroom

Because such a large percentage of water use is in the bathroom, this is where water conservation efforts should begin. Here are some key tips:

- Install water-saving devices. Devices for faucets and shower heads are depicted in the accompanying illustrations.
- Take shorter showers. Unless a shower lasts seven minutes or less, bathing in the tub will use less water and will conserve energy. A kitchen timer is a useful bathroom accessory. The "Navy" shower—use water to get wet, turn off water while lathering, turn on water to rinse—uses the least water. The amount of water used in a tub-shower is easily determined by closing the drain during the shower.

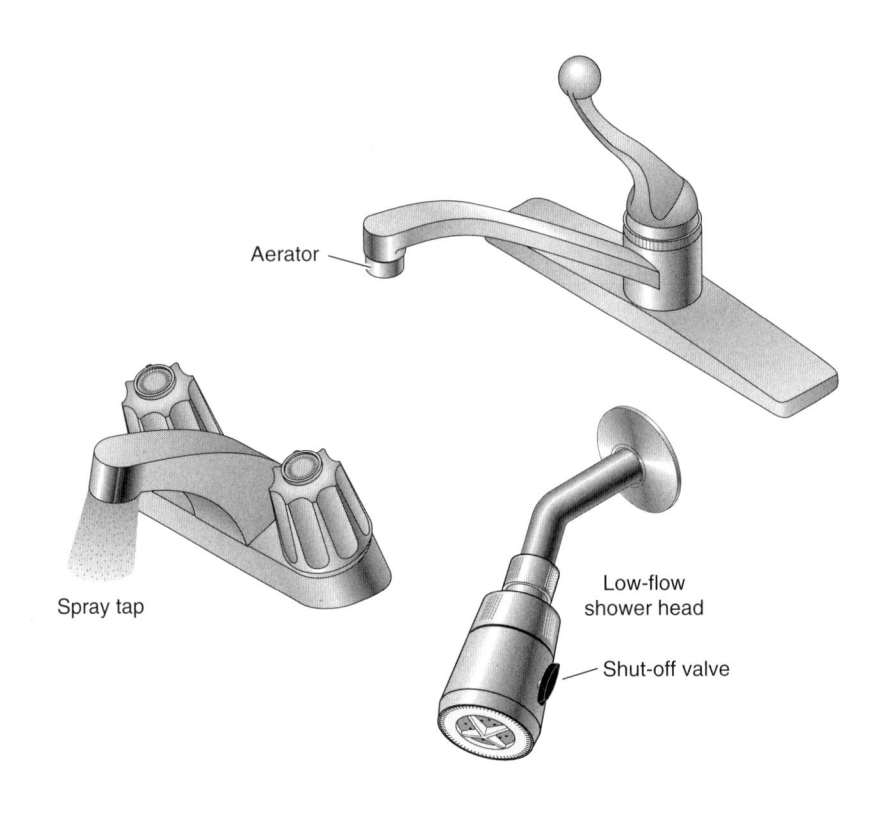

Water-saving devices

- **Aerators** *reduce splashing. They also introduce bubbles into the stream of water, making the volume of water flowing from the faucet seem greater than it actually is.*
- **Spray taps** *deliver water in a broad pattern like a small shower head. This makes a low flow of water more efficient for rinsing.*
- **Shut-off valves** *allow you to turn off the water while soaping or shampooing and then turn the water back on.*
- **Low-flow shower heads,** *now required in new construction, deliver anywhere from about 1½ to 2½ gallons of water per minute. Older style shower heads—the kind still in many homes—deliver 5 to 7 gallons per minute.*

Rock-filled plastic container

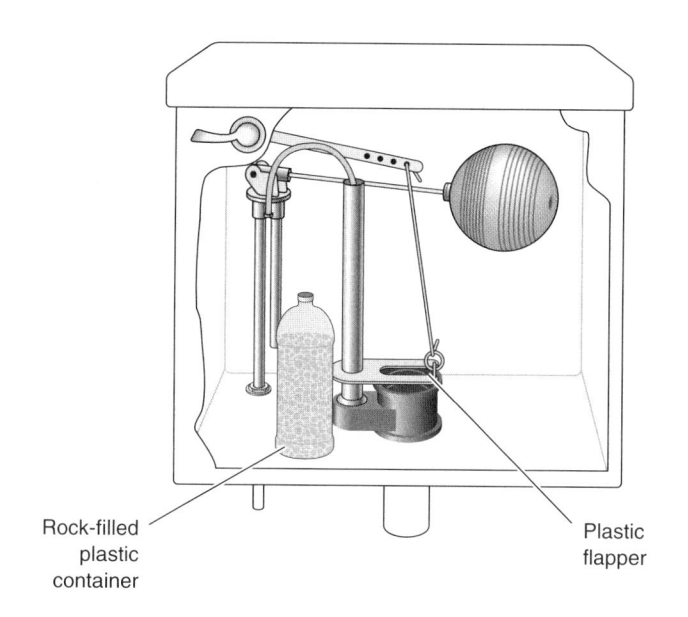

Plastic flapper

Placing a rock-filled container in the tank of a conventional flush toilet can reduce the amount of water used by up to 5 to 10 percent.

- When filling the tub, don't let water run down the drain while waiting for it to get hot. Instead, close the drain before turning on the faucet. The water will soon be hot and the temperature can be adjusted as the tub fills.
- Place "toilet dams" or rock-filled containers in the tanks of conventional flush toilets. Rock-filled containers can reduce the amount of water used by 5 to 10 percent, while toilet dams can reduce it by up to 25 percent—without affecting flushing ability. However, never use a brick to accomplish the same effect because particles from it could harm your plumbing.
- Consider replacing current toilets with the new "low-volume" toilets now on the market. Some use only a gallon of water per flush (compared to 5 gallons per flush for older toilets).
- Don't use the toilet to just flush away facial tissues, paper, and other similar solid and liquid wastes. Using a wastepaper basket is a lot cheaper.
- Repair leaks in your faucets and toilets. A leaky faucet can waste 20 gallons or more per day, whereas leaky toilets can waste hundreds of gallons per day. Because a toilet leak goes directly down the drain, it often goes unnoticed. To find out if your toilet leaks, put a little food coloring in the tank. If, without flushing, color appears in the bowl,

you have a leak that should be repaired. Leaky toilets often can be repaired by adjusting the float arm or replacing the flush ball or flapper valve. Repairing a leaky faucet is usually as simple as changing an inexpensive washer.

- Turn off the water while brushing your teeth; this step can save a family 5 to 10 gallons per day (or 3,650 gallons a year).
- Rinse hand razors in a filled sink rather than under running water.
- When shampooing, turn off the water while lathering the hair.

Conserving water in the kitchen

- Do not let faucets run for washing or rinsing. Always fill a container with water for this purpose, or use the sink by stoppering the drain.
- Use a brush, washcloth, or your hand to dislodge particles of dirt when washing anything rather than relying on the force of the water to do the job.
- When filling a kettle, try to estimate the actual amount needed; leaving unused boiled water on the stove means that both water and energy have been wasted.
- Use plastic ice trays. That way, you can loosen the cubes by twisting the tray rather than by running water over it.
- Keep a bottle of water in the refrigerator rather than letting water run in the sink to get a cool drink.
- When cleaning vegetables, use a filled sink and a vegetable brush, and wash all of the vegetables for a meal at the same time.
- Use only the minimum amount of water to cook foods; otherwise, flavor and food value may be wasted along with water.
- Do not use garbage disposals except at the end of cooking or cleanup periods or when full. Whenever possible, don't use the disposal at all; compost vegetable peelings for your garden, or put them in the garbage can.
- Run an automatic dishwasher only with full loads. Do not prewash dishes unless necessary. To save energy, use the no-heat drying cycle, or turn the dishwasher off at the start of the dry cycle, open the door, and let the dishes air-dry.
- Do not use the extra-long prewash and scrub cycles on the dishwasher unless absolutely necessary.

Conserving water outdoors

- Water only when necessary. It takes 660 gallons of water to supply 1,000 square feet of lawn with 1 inch of water. This is nearly the same amount of water that you use inside the house in an entire week.

To maintain green, actively growing turfgrass throughout the entire season, most lawns require 1 to 1½ inches of water per week—no matter whether the water is coming from rain or irrigation. However, if maintaining a green lawn all summer long is not important to you, or is too expensive, or if water use is restricted in your area, you may want to stop watering and let the turf go dormant. Allowing the lawn to go dormant normally will not hurt the grass, except under extended periods of extreme heat. But be aware that when you allow it to go dormant, appearance suffers and weeds can increase.

- Water the lawn in the early morning to avoid evaporation losses.
- For successful lawn irrigation, irrigate slowly, deeply, and infrequently. In the flower or vegetable garden, use drip irrigation or soaker hoses to water because they apply water slowly and directly to the soil. This helps ensure that all of the water remains on-target. Also, because the foliage stays dry, the incidence of disease is reduced.
- Avoid sprinklers that produce a fine mist; too much water is lost by wind and evaporation.
- Use an alarm clock or the stove timer to remind you to shut off the sprinklers.
- Use pistol-grip nozzles (spring shutoff) on all hoses to avoid waste, and always turn off the faucet tightly to prevent leakage.
- Reduce evaporation losses from flower and vegetable gardens by using an organic mulch or plastic ground cover between rows.
- Sweep sidewalks and driveways instead of washing them down with the hose.
- Collect water from roof gutters to use for lawn and plant watering.
- Do not water lawns or wash cars when water is in short supply.
- When washing the car, rinse it once, use a bucket of soapy water to wash it, and then give it one more quick rinse. Taking it to a car wash may save water because many commercial installations recycle their water.
- Leave grass clippings in place. Clippings provide a kind of mulch, which reduces the evaporation of water in the soil and insulates the soil from the heat of summer.

> **CHECK IT OUT**
> For details on using clippings as mulch: page 162.

Other tips

- Run clothes washers only with a full load unless a reduced fill setting is available. Use "warm" or "cold" settings if possible.
- When you are purchasing a new clothes washer, choose one with variable load or suds-saver options.
- When you go on an extended vacation, turn off the water to the house; a leak while you are away could be expensive and could do a lot of damage. Be sure, however, to turn off the water heater also. If it should begin to leak and drain dry, it could burn out.
- Locate the water heater near the points of most hot water use. Consider a separate water heater for distant bathrooms.

Prepared with Henry Spies, UI School of Architecture—Building Research Council; and Mel Bromberg, drinking water and health specialist, UI Cooperative Extension Service.

For more information

U.S. Environmental Protection Agency

PUBLIC INFORMATION CENTER
401 M St., SW
Washington, DC 20460
(202)260-7751

Regional EPA offices

REGION 1
Boston, MA
(617)565-3420

REGION 2
New York, NY
(212)637-3000

REGION 3
Philadelphia, PA
(215)597-8320

REGION 4
Atlanta, GA
(404)347-4727

REGION 5
Chicago, IL
(312)353-2205

REGION 6
Dallas, TX
(214)665-6444

REGION 7
Kansas City, KS
(913)551-7000

REGION 8
Denver, CO
(303)293-1603
(800)227-8917

REGION 9
San Francisco, CA
(415)744-1305

REGION 10
Seattle, WA
(206)553-1200
(800)424-4372

NOTE: Every state has its own environmental agency. Your regional EPA office should be able to tell you how to contact your state agency.

Appliances

AMERICAN COUNCIL FOR AN ENERGY-EFFICIENT ECONOMY
2140 Shattuck Ave., Suite 202
Berkeley, CA 94704
(510)549-9914

Offers publications and answers questions about home energy conservation and energy-efficient appliances.

WOOD HEATER PROGRAM
Management, Energy, and Transportation Division (2223A)
U.S. Environmental Protection Agency
401 M Street SW
Washington, DC 20460
(202)564-7021

Backyard wildlife

BACKYARD WILDLIFE HABITAT PROGRAM
National Wildlife Federation
8925 Leesburg Pike
Vienna, VA 22184-0001
(703)790-4000

Operates 7 a.m. to 3:30 p.m., Central Time. Provides information on how to create and certify your own backyard wildlife habitat.

Compact fluorescent light bulbs

THE GREEN LIGHTS HOTLINE
U.S. Environmental Protection Agency
(202)775-6650

Consumer safety

U.S. CONSUMER PRODUCT SAFETY COMMISSION
Washington, DC 20207-0001
Product Safety Hotline: (800)638-CPSC

Recorded information is available 24 hours a day (in English or Spanish) when calling from a touch-tone phone. Operators are on duty Monday through Friday from 9:30 a.m. to 3 p.m., Central Time, to take complaints about unsafe consumer products.

Drinking water

SAFE DRINKING WATER HOTLINE
(800)426-4791

Operates Monday through Friday, 8 a.m. to 4:30 p.m., Central Time. Provides information on drinking-water contaminants, including possible health effects and contamination sources. Also makes referrals for water-testing laboratories in your area.

Energy conservation

ENERGY EFFICIENCY AND RENEWABLE ENERGY CLEARINGHOUSE
P.O. Box 3048
Merrifield, VA 22116
(800)363-3732

Operates Monday through Friday, 9 a.m. to 7 p.m., Central Time. Provides consumer information on conservation and renewable energy in residences.

Indoor air quality

AMERICAN LUNG ASSOCIATION
1740 Broadway
New York, NY 10019-4374
(800)LUNG-USA

This number automatically puts you in contact with the nearest office of the American Lung Association.

INDOOR AIR QUALITY INFORMATION CLEARINGHOUSE
P.O. Box 37133
Washington, DC 20013-7133
(800)438-4318
(202)484-1307
FAX: (202)484-1510

*Operates Monday through Friday from 8 a.m. to 4 p.m., Central Time.
Distributes EPA publications, answers questions on the phone, and makes referrals to other nonprofit and governmental organizations.*

Lead

NATIONAL LEAD INFORMATION CENTER
(800)LEAD-FYI

Operates 24 hours a day, seven days a week. Callers can order an information package by leaving name and address on a recorded message. Message is in English or Spanish. To speak with an information specialist, call (800)424-LEAD. Information specialists are on duty Monday through Friday, 7:30 a.m. to 4 p.m., Central Time.

Pesticides

NATIONAL PESTICIDE TELECOMMUNICATIONS NETWORK
Agricultural Chemistry Extension
Oregon State University
333 Weniger Hall
Corvallis, OR 97331-6502
(800)858-PEST
FAX: (503)737-0761

*Operates Monday through Friday, 8:30 a.m. to 6:30 p.m., Central Time.
Provides information about pesticides to the general public and to the medical, veterinary, and professional communities.*

Radon

NATIONAL RADON HOTLINE
(800)SOS-RADON

Operates 24 hours a day. Provides recorded information. Leave your name and address, and information will be sent to you—including your local radon contact.

Trees

INTERNATIONAL SOCIETY OF ARBORICULTURE
P.O. Box GG, 6 Dunlap Ct.
Savoy, IL 61874
(217)355-9411

Produces educational materials and maintains a computerized information service. The International Society of Arboriculture also can give you a list of certified arborists where you live.

NATIONAL ARBOR DAY FOUNDATION
211 N. 12th St.
Lincoln, NE 68508
(402)474-5655
FAX: (402)474-0820

Sponsors National Arbor Day, Tree City USA, Conservation Trees programs, and workshops on various urban forestry topics. Also publishes newsletters and an Arbor Day Kit for fifth graders.

Waste reduction and recycling

EPA SOLID WASTE HOTLINE
(800)424-9346
In Washington, DC: (703)920-9810

Operates Monday through Friday, 8 a.m. to 5 p.m., Central Time. Can help you to contact recycling coordinators in your area. Also provides information on EPA regulations.

NATIONAL RECYCLING COALITION
1727 King St., Suite 105
Alexandria, VA 22314
(703)683-9025

Provides information and publications on waste reduction and recycling.

WASTE WATCH CENTER
16 Haverhill St.
Andover, MA 01810
(508)470-3044
FAX: (508)470-3384

Can help you locate the nearest household hazardous-waste collection site.

Additional sources

CONSUMER INFORMATION CENTER
Pueblo, CO 81009

Publishes the Consumer Information Catalog, *which lists more than 200 free and low-cost federal publications. When requesting the catalog, write to:* Consumer Information Catalog *(then the address above). You can also order the catalog by calling: (719)948-4000.*

The Consumer Information Center also publishes the Consumer's Resource Handbook, *which provides consumer tips on everything from home improvement to product safety. When writing for the handbook, write to:* Handbook *(then the address above).*

NATIONAL CANCER INSTITUTE
Building 31, Room 10A07
9000 Rockville Pike
Bethesda, MD 20892-2580
(800)4-CANCER

Offers free publications.

NATIONAL HEART, LUNG, AND BLOOD INSTITUTE
Information Center
P.O. Box 30105
Bethesda, MD 20824-0105
(301)251-1222

Operates Monday through Friday, 7:30 a.m. to 4 p.m., Central Time.

Home environment goals

Check your progress

57 Ways to Protect Your Home Environment actually includes a lot more than 57 things you can do around the home. With this amount of information, there's always the risk of feeling overwhelmed—unless you set realistic goals and follow up on them. This checklist can help you do just that. Here's how:

- Scan the list and check as "completed" any of the things you have already done.
- Of the tasks you haven't done yet, put a check by the ones you would like to do.
- Whenever you accomplish one of the goals, mark a check or write the date in the "Completed" column.

A lot of the tasks in this checklist are things that you may only need to do once; however, certain tasks may need to be repeated on a somewhat regular basis. If your drinking water comes from a well, for instance, you will want to test for bacteria and nitrate annually. Also, not everything on this checklist will apply to you. As one example, your furnace may not be old enough to justify replacing it with a high-efficiency one. Read the appropriate chapters to find out what applies and what doesn't.

Checklist of home improvement goals

Caring for the home landscape

	To do	Completed
Reduce the number of times you fertilize the lawn per year (Chapter 1).		
Plant a low-maintenance lawn (Chapter 1).		
Have the soil in your garden and/or yard tested for fertility levels (Chapters 1 and 3).		
Purchase an electric lawn mower (Chapter 2).		
Purchase a reel mower (Chapter 2).		
Rotate crops in the garden (Chapter 3).		
Select crops that require fewer pesticides (Chapter 4).		
Plant native trees and/or shrubs in the yard (Chapter 5).		
Check the health of trees in your yard (Chapter 6).		
Plant a woodland and/or prairie garden (Chapter 7).		
Replace your lawn with native grasses and wildflowers (Chapter 7).		
Install bird feeder(s) (Chapter 8).		
Select flowering plants that attract butterflies (Chapter 8).		

Using alternatives to traditional pesticides and fertilizers—where practical

Scout for pests in the yard and garden (Chapter 9).		
Use microbial insecticides (Chapter 10).		
Use botanical insecticides (Chapter 11).		
Use insecticidal soaps (Chapter 11).		
Take steps to conserve beneficial insects (Chapter 12).		
Purchase traps to monitor or control insects (Chapter 13).		
Begin to use pest barriers in the garden (Chapter 14).		
Try solarizing soil (Chapter 15).		
Use an organic fertilizer in your garden (Chapter 16).		

Using chemicals on the landscape—if necessary

Ask your lawn-care company about less toxic alternatives (Chapter 17).		
Read the labels of the pesticides you own and buy (Chapter 19).		
Purchase protective equipment for mixing and applying pesticides (Chapter 20).		
Calibrate your pesticide application equipment (page 112).		
Calibrate your fertilizer application equipment (page 112).		

Checklist of home improvement goals, cont.

Storing and disposing of hazardous chemicals

To do   *Completed*

Store hazardous materials in a locked cabinet out of reach of children (Chapter 22).
Place all flammable materials in a detached garage or outside storage shed (Chapter 22).
Take an inventory of household hazardous materials (Chapter 22).
Dispose of household hazardous materials that you no longer need (Chapters 22-26).
Switch to less toxic antifreeze (Chapter 24).
Begin to recycle used oil and used oil filters (Chapter 24).
Try solvent-free paint (page 130).

Managing yard waste, food waste, and other household waste

Limit junk mail by writing to the appropriate services (Chapter 27).
Plan ways to reduce the amount of waste you generate (Chapter 27).
Contact the local recycling center for a complete list of the materials they accept (Chapter 28).
Reorganize your system for collecting recyclable materials (Chapter 28).
Begin to recycle paper (Chapter 28).
Begin to recycle metal (Chapter 28).
Begin to recycle glass (Chapter 28).
Begin to recycle plastic (Chapter 28).
Find a spot where you can collect bags and other items to reuse (Chapter 29).
Go through your home and collect everything that you can give to others for reuse (Chapter 29).
Evaluate the environmental impacts of the products you buy (Chapter 29).
Begin to leave grass clippings on the lawn (Chapter 30).
Make or obtain a compost bin (Chapter 31).
Use yard waste as mulch (Chapter 32).

Protecting your drinking water

Test your drinking water for nitrate and bacteria (Chapters 34-35).
Test your drinking water for pesticides (Chapters 34 and 36).
Test your drinking water for other contaminants (Chapters 34 and 37).
Install a water-treatment system (Chapter 38).

Protecting your drinking water, cont.

To do   Completed

Have a reputable contractor pump out your septic system (Chapter 39).
Compare your tap water with bottled water (Chapter 40).
If chlorine taste is a problem, take steps to remedy it (Chapter 40).

Protecting the indoor environment

Inspect your home for sources of lead (Chapter 41).
Test your children for lead levels in their blood (Chapter 41).
Test for radon (Chapter 42).
Install a radon mitigation system (Chapter 43).
Install an air filter to control dust (Chapter 44).
Purchase mattress and pillow covers that block mites (Chapter 44).
Install a humidity gauge to maintain proper humidity levels in summer and winter (Chapter 44).
Install glass doors on fireplaces (Chapter 45).
Inspect your wood-burning stove for proper installation (Chapter 45).
Install carbon monoxide detector(s) (page 225).
Check your home for formaldehyde sources (Chapter 46).
Take steps to reduce formaldehyde exposure (Chapter 46).
Check your home for asbestos sources (Chapter 46).
Use less toxic alternatives to household pesticides (Chapters 47 and 49).
Switch to liquid dishwashing detergents, laundry detergents, and bleaches (Chapter 50).

Conserving energy and water

Turn the thermostat down 10 degrees at night during winter (Chapter 52).
Close unoccupied rooms and partially close registers to keep these rooms cooler (Chapter 52).
During summer, close air registers in the basement (Chapter 52).
Shade windows (Chapters 52 and 56).
Install a high-efficiency furnace (Chapters 52 and 54).
Install a high-efficiency air conditioner (Chapters 52 and 54).
Caulk and weatherstrip the house where needed (Chapter 53).
Add insulation where needed (Chapter 53).
Install better-insulated windows (Chapter 53).
Purchase an energy-efficient refrigerator (Chapter 54).

Checklist of home improvement goals, cont.

Conserving energy and water, cont.

	To do	Completed
Purchase an energy-efficient dryer (Chapter 54).	___	___
Use compact fluorescent light bulbs (Chapter 55).	___	___
Plant trees and shrubs to shade the west wall and either the south end of the east wall or the east end of the south wall (Chapter 56).	___	___
Use shrubs, trees, or a fence to prevent heat loss or gain from windows and entryways (Chapter 56).	___	___
Check your home for leaky faucets and toilets; then repair them (Chapter 57).	___	___
Install a low-flow shower head (Chapter 57).	___	___
Install aerators on faucets (Chapter 57).	___	___
Purchase a low-volume toilet (Chapter 57).	___	___
Place a rock-filled container or "toilet dam" in the toilet tank (Chapter 57).	___	___
Determine whether showers or baths use less water (Chapter 57).	___	___
Purchase a drip irrigation hose for watering the garden (Chapter 57).	___	___

Water test results

Contaminant	Date of test	Level detected	Maximum contaminant level (MCL)[a]
Nitrate			44 milligrams per liter (mg/l)
Nitrate-nitrogen			10 mg/l
Bacteria			Any bacteria detected is a potential problem.
Lead			15 parts per billion (ppb)[b]

[a] An MCL is the maximum contaminant level allowed in a public water supply.
[b] No amount of lead in water is good. Some specialists recommend that people take action to solve a lead problem when levels reach 5 ppb.

Blood-lead levels

Lead levels greater than 10 deciliters per microgram indicate the need to check for sources of lead in the home. Do not panic. Medical treatments and therapies are available for reducing lead in the bloodstream, but a physician or local public health nurse should advise you on these.

Name of family member	Blood-lead level	Date of test	Name of lab or hospital

Radon test results

The U.S. EPA guideline for radon is 4 picoCuries per liter (pCi/L). However, some radon tests express the results as "working levels" (WL). 0.02 working levels is the equivalent of 4 pCi/L.

Date of test	Level detected
_____	_____
_____	_____
_____	_____
_____	_____
_____	_____
_____	_____

Notes on inspections, other test results, etc.:

Index

Note: Entries in italic type indicate charts or illustrations.

Absorption system, 196
Adhesives, 131
Aerators, 284, *284*
Aerosol containers, 131–2
Aerosols, 241
Air cleaners, 217
Air conditioning. *See* Cooling
Air ducts, 257
Air quality, 246, 265, 292
　See also Pollution; Ventilation
Air stripping, 189–90, *192*
Air-to-air heat exchanger, 256
Alcohol, 248
Algae, 218
Allergens, 216–20
Allergies, 216
American Council for an Energy-Efficient Economy, 290
American Lung Association, 292
Ammonia, 248
Animal dander, 219
Anion exchange, 190, *192*
Ant, 234, *236, 239*
Ant baits, 241
Antifreeze, 123–4
Apples, 20
Appliances, 133, 266–71, 290
　disposal of, 133
　energy costs, 267–70
　labels, 266–*7, 268–9,* 270
　maintaining, 270–1
Arborists, 87–9
Asbestos, 229–30
Asthma, 216
Attics, 251
Auto products, 123–5, 145

Bacillus thuringiensis, 57, *60*
Back drafts, 224, 226, 257
Backyard Wildlife Habitat Program, 291
Bacteria, insecticides, 57, *60*
Bacteria, as a contaminant, 174–5, 188–9, *192,* 193, 200, *301*
Baited traps, 73–4
Baits, 240–1, 245
Baking soda, 248
Bald-faced hornet, *238*
Ballasts, 273, *274,* 275
Bark, as mulch, 164
Bathrooms, water use and conservation, 283–6
Bats, 47–8
Batt insulation, 258

Batteries, 124, 134–5, 151
Bean weevils, *235*
Beetles, 61, *68, 69, 235*
Big-eyed bug, *68*
Bird feeders, *42,* 42–5
Birds, 41–5
　foods for, 43–4
　water for, 45
Block-wall ventilation, 214, *215*
Blue-baby syndrome, 176–7
Boilers, 221–2
Botanical insecticides, 10, 62–6, 244–5
Bottled water, 198–200
　regulation, 199–200
　sources of, 201
　types of, 201
　versus tap water, 198–9
Boxelder bug, *236*
Bt, 57, 60, *61,* 67
Buckwheat hulls, as mulch, 164
Bug zappers, 76
Butterflies, 45, *46,* 51–2
　food for, *46*

Calibration, 104, 109, 112–3
Cancer, 180, 208
Carbon filters, 188–9, *189, 191–2*
Carbon monoxide, 224–5, 265
Carbon monoxide detectors, 225
Carpenter ant, 234, *239*
Carpenter bee, *239*
Carpet beetle, *235*
Cation exchange, 190, *192*
Caulking, 258–9
CCA-treated wood, 32, 222
Centipede, *237*
Chemical communication, in insects, 73
Chemical reactions, 66
Children, blood-lead levels in, 207
Chimneys, 221-2, 251
Chinese praying mantid, *68*
Chipmunks, 48
Chlorination, 184, *192,* 193
　chlorine taste, 199
Christmas trees, 166–7
Closets, 219
Clover mite, *237*
Cluster fly, *237*
Cockroaches, 218, 234, *235, 236*
　control of, 218, 241, 244
Cocoa-bean hulls, as mulch, 164
Coliform bacteria, 174–5, *192*

Color rendition index (CRI), 277
Combustion equipment, 221–6
Common green lacewing, *68*
Compact fluorescent light bulbs, 1–2, 272–8, *273–7,* 291
  ballasts, 273, *274,* 275
  color rendition index (CRI), 277
  costs, *273*
  design, 273
  emission reductions, *277*
  places to buy, 278
  style, 275, *276*
Companion plants, 72
Compost, 18, 155–61, 163
  high-carbon materials, *159*
  high-nitrogen materials, *159*
  ingredients, 158, *159*
  as mulch, 163
  trouble shooting, *159*
  types of bins, 160–1, *160–1*
Consumer Information Center, 294
Consumer Product Safety Commission, 291
Consumers, household cleaners, 247
Contaminants
  asbestos, 229–30
  clues in drinking water, *186–7*
  from combustion equipment, 221–6
  formaldehyde, 227–8
  lead, 204–7
  in septic system, 196–7
  radon, 188–9, 208–15, 255, 292, 302
  See also specific items
Cooling, 252–4
Copper sheeting, 78
Core aerification, 7–8
Corncobs, as mulch, 164
Corrosive materials, 94, 117
Cosmetics, 132
Covering exposed earth, 214, *215*
Crawl space, 208, *209,* 214, 251, *259*
Creosote, 32
Crops
  pest control, 15–8, 19–20, *19*
  resistant varieties, 20
  rotation, 16

Damsel bug, *68*
Deciduous shrubs, *23–6*
Deciduous trees, *22–3*
Deer mice, 48
Des Moines Water Works, 182–5
Diapers, 153
Diatomaceous earth, 244
Disposal
  of adhesives, 131
  of aerosol containers, 131–2
  of appliances, 133
  of auto products, 123–5
  of batteries, 124, 134–5
  of cosmetics, 132
  of household cleaners, 132
  of medications, 132
  of paints and solvents, 128–30
  of pesticides, 121–2
  of tree residues, 166–7
Distillation, 189, *192*
Doors, weatherstripping, *260*
Drain fly, *235*
Drain-tile suction, 212, *215*
Drinking water. *See* Water, drinking
Drop spreader, *108,* 108–9
Drought tolerance, 21, *22–6*
Dryers, 270
  See also Appliances
Dust, 216–20
Dust mites, *216,* 216–20
Dusters, 103

Earthworms, 8
Earwig, *237*
Electric mowers, *12–3,* 12–4
Electrostatic filters, 217
Electrostatic precipitators, 216–7
Energy efficiency, 299–300
  appliances, 266–71
  in heating and cooling, 252–7
  in house design, 250–1, 255–7
  landscaping, 279–82
  light bulbs, 272–8, *273–7*
Energy Efficiency and Renewable Energy Clearinghouse, 291
EnergyGuide labels, 266–70, *268–9*
Environmental Protection Agency, 180
  address of, 290
Environmental trade-offs, 153
EPA Solid Waste Hotline, 293
Estate sprayers, 102
Evergreen shrubs, *26*
Evergreen trees, *23*
Evergreens, 281–2
Explosive materials, 118

Fertilizer and fertilization
  application, 27–8, 108–13, *108–11*
  calibration, 109, 112–3
  for gardens, 16
  for lawns, 5, *6*
  natural, 81–3, *297*
  nitrogen applications, 5, *6,* 82
  phosphorus applications, 83
  potassium applications, 83
  surface broadcasting, 27

Fertilizer and fertilization (cont.)
  for trees, 27–8, *28*
Fescues, *6*
  desired height, *8*
Fiber board, *227*
Field cricket, *237*
Filtering water. *See* Water treatment
Fine-leaf fescues, *6*
  desired height, *8*
Fireplaces, 222, 226, 257
Flammable materials, 118, 119–20
Flea, *235,* 243
Flour beetle, *235*
Fluorescent light bulbs. *See* Compact fluorescent light bulbs
Fluoridation, 184
Fly traps, 75
Foam boards, 258
Foam disks, *77, 78*
Food
  to absorb lead, 207
  lead in, 206–7
Food waste, 143
Formaldehyde, 227–8, 232
  sources of, 227
Fruits
  disease-resistant, 20
  insect-resistant, 20
  pest control, 19–20
Fungi, 58, 218
  as insecticide, 58
Furnaces, 222, 253, 254, 256

Garbage-can compost bin, *160*
Gardens, 15–20, *19*
  air circulation, 17
  crop residue, 18
  crop varieties, 15
  location, 15
  mulch for, 162–5
  pest barriers, *77,* 77–8
  planting dates, 17
  rotating crops, 16
  seeds, 15
  soil, 16–7
  solarizing soil, 79–80
  watering, 18, 286–7
  See also Fertilizer and fertilization; Pests
German cockroach, *235*
Glass, 142–3
Granular applicators, 103
Grass, 4–14
  common grasses, *6*
  desired heights, *8*
  selection, 5
  native grasses, 36, *38–40*
  See also Lawn

Grass clippings, 5, 154–5, 156, 158, *159,* 163
Grasshopper control, 58, *60*
Green lacewing, *68*
Green Lights Hotline, 277, 291
Greensand filter, 193
Ground beetle, *68*
Groundhogs, 50

Habitat modification, 243
Hardiness zones, 33, *34*
Hazardous materials, 116–20, 126–7, 298, *117–8*
  storing, 119
Hazardous waste
  disposal of, 126–7, 131–2
  See also specific item
HDPE plastics, 143, *144*
Head louse, *236*
Health advisories, 180
Heat recovery ventilation, 212–3, *215,* 253–4, 256
Heating, 252–7
Heavy metals, 246–7
HEPA filters, 217–8
Herbal insecticides, 64
Home Environment Assessment List, 231–2, *232–3*
Home improvement goals, 297–302
Honey bee, *238*
Hops, as mulch, 164
Hose-end sprayers, 101, *101*
House design, 250–1, 255–7
Household cleaners, 66, 132, 246–8
  natural, 247–8
Household products
  as hazardous materials, 116, 126–7
  household cleaners, 66, 132, 246–8
  as pesticides, 66
  See also specific contaminants
Household waste reduction, 138–40, 298
Humidifiers, 220
Hybrids, 20
Hydrogen peroxide, 248

Illinois, 149–50
Indianmeal moth, *236*
Indoor Air Quality Information Clearinghouse, 292
Indoor environment. *See* Contaminants
Insect traps, 73–6, *74,* 245
Insecticidal soaps, 62, 65, 244–5
Insecticides, 56–66, 95, 244–5, *60*
  botanical, 10, 62–6, 244–5
  insecticidal dusts, 244
  insecticidal soaps, 10, 62, 65, 244–5
  microbial, 10, 56–9, *60,* 61
  See also Pesticides

Insects, 234–9, *235–9*
  beneficial, 67–71, *68–9*
  color, 75
  household, 234–9, *235–9*
  See also Insecticides; Pesticides; Pests; *specific insects*
Insulation, 251, 255, 258–65, *259*, 279
  R-values, *259*
  for recommended areas in the house, *259*
  U-values, 263, *263*
  windows, 263, *264*, 265
Integrated pest management, 4, 11, 87
International Society of Arboriculture, 293
IPM. *See* Integrated pest management

Junk mail, 138

Kentucky bluegrass, *6*
  desired height, *8*
Kerosene heaters, 223
Kitchens, water conservation in, 286

Lacewing, *68*
Lady beetle, *68, 69*
Landfill, 153, 155
Landscaping, 279–82, *280, 281,* 297
  See also Lawn
Lawn, 4–14, *6–8, 12–3*
  core aerification, 7–8
  fertilization, 5, *6*
  nitrogen applications, 5, *6*
  pests, 9–11
  in shady areas, 4
  site evaluation, 4
  soil preparation, 4–5
  turfgrass selection, 5
  watering, 7, 9, 286–7
  weeds, 9
Lawn-care companies, 86–7
  toxins, 90–1
LD$_{50}$ values, 92
Lead, 204–7, 231–2, *232–3*, 292
  blood-lead levels, 207, 232, 301
Leaves
  burning, 167
  as compost, 167
  as mulch, 163, 167
Lemon juice, 248
Light, insects, 75
Light bulbs. *See* Compact fluorescent light bulbs
Limonene, 64
Linalool, 64
Live traps, 47–50
Loose-fill insulation, 258
Lumber, safety of treated wood, 32, 222

Lung cancer, 208
Lye, 248

Manure
  as fertilizer, 81–3, 156
  as mulch, 165
Mechanical dethatching, 8
Mechanical filtration, 190, *192*
Medications, 132
Metals, 142
Mice, 48, *237*
Microbial insecticides, 10, 56–61, *60*
Midwest
  native prairie plants, *38–40*
  native trees, *22–6*
Mildew, 218
Millipede, *237*
Minerals, in drinking water, 172
Minneapolis, Minnesota, 255–7
Minnesota's Cold Climate Housing Project, 255–7
Mites, *69, 216,* 216–20
Mold, 218, 220
Moles, 48
Motor oil, 123, 145
Mowing grass, 8, *8,* 12–4, *12– 3*
Mulch, 155, 162–5
Mulching mower, 5, 154–5

National Arbor Day Foundation, 293
National Cancer Institute, 294
National Heart, Lung, and Blood Institute, 294
National Lead Information Center, 292
National Pesticide Telecommunications Network, 292
National Radon Hotline, 292
National Recycling Coalition, 293
Native plants, 33–40
  fall blooming, *40*
  late-spring blooming, *38*
  in Midwest, *38–40*
  shrubs, 21, *23–6*
  summer blooming, *38–9*
  trees, 21, *22–3*
  where to find, 35–6
Natural gardens, 33–40, *38–40*
Neem, 64
Nematodes, 58–9, *60,* 61
Nicotine, 64, 66
Nitrate, 172, 174–7, *175*
  removal, 177, 184, 189–90, *192*
Nitrogen, 5, *6,* 82, 158–9, *159*
Northeast, native trees and shrubs, *22–6*
Nosema locustae, 58

Oil filters, 123
Organic pesticides. *See* Botanical insecticides; Insecticidal soaps; Microbial insecticides
Organic fertilizers, 81–3, 91, 156
Oriental cockroach, *236*
Ovens, 223
Oxalic acid, 248
Oxidizing filter, *192,* 193
Ozonation, *192,* 193

Packaging, 139–40
PAHs, 221, 222
Paint thinner, 129
Paints, 128–30
   lead, 204–5
   solvent-free, 130
Paper, 142
Paper versus plastic bags, 153
Paper wasp, *238*
Parasites, 67
Particle board, *227,* 228
Passive electrostatic filters, 217
Peanut hulls, as mulch, 165
Peat moss, as mulch, 165
Pecan shells, as mulch, 165
Pentachlorophenol, 32
Perennial ryegrass, *6*
   desired height, *8*
Pest barriers, *77,* 77–8
Pest-control operators, 234–5
Pesticide containers, 121–2
Pesticide Telecommunications Network, 179
Pesticides, 56–61, *60,* 292, 297
   alternatives to, 9–11, 56–9, 62–5, 242–5, 297
   application of, 101–7, *101-3*
   backyard prairies, 37–8
   calibration, 104
   chemical reactions, 66
   disposal of, 121–2
   in drinking water, 172, 178–81, 188–90, *192,*193
   formulations of, 95–6
   for fruits and vegetables, 19–20
   homemade, 66
   household, 240–1
   household products as, 66
   IPM, 4
   labels, 94, 97–100, *98–100*
   microbial insecticides, 56–61, *60*
   mixing, 105
   names of, 95
   overuse, 121
   residual, 240
   spot-treating problems, 10, 54, 90
   toxins, 66, 90–1, 92–6
   types of, 95

Pests
   common household, 234–9, *235–9*
   companion plants, 72
   controlling, 243
   exclusion, 242
   habitat modification, 243
   monitoring, 73
   scouting for, 54–5
   understanding, 242
PET plastics, 143, *144*
pH, 17, 81, 163
Phosphates, 246
Phosphorus, 83
Pine needles, as mulch, 165
Pit composting, 161
Plant covers, 77, *77*
Plant oils, 64
Plants
   companion, 72
   diversity, 35
   for energy conservation, 279–82, *280–1*
   hardiness zones, 33, *34*
   mulch for, 162
   native, 33–40, *38–40*
   selection, 33–5, *34*
   for shade, 279–80, *280*
   for windbreaks, 279, *281,* 281–2
Plastics, 1, 143–4
   codes, *144*
Pleated filters, 216–7, *217*
Plywood paneling, *227*
Pollen, 219
Pollution. *See* Contaminants
Porches, 251
Potassium, 83
Powder-post beetle, *239*
Prairie, 33–40, *38–40*
Praying mantid, 67, *68*
Predators, 67
Protective clothing, *103,* 104–5
Protozoa, 58, *60*
Protozoan pathogens, 58
Pyrethrins, 63
Pyrethrum, 63

Rabbits, 49
Raccoons, 49
Radon, 208–15, *209, 213, 215,* 255, 292
   in drinking water, 188–9, *192,* 193
   reduction of, 212–5, *213, 215*
   sources, *209*
   testing, 210–1, 302
Ranges, 223
Reactive materials, 118
Reclamation programs, 166–7
Recycling, 1, 141–6, *141, 144, 145,* 293, *298*

Recycling (cont.)
  auto products, 123–5, 134
  batteries, 134–5
  centers, 149–50
  contaminants in the process, 147–8
  glass, 142–3
  grass clippings, 154–5
  metals, 142
  paper, 142
  plastics, 143–4, *144*
  symbols, *144–5*
Reduce, recycle, reuse, 152
Reel mowers, 14
Refrigerators
  maintaining, 270–1
  See also Appliances
Reusing, 151–3
Reverse osmosis, 190, *191–2,*
Roach baits, 241
Roaches. *See* Cockroaches
Rodent baits, 240–1
Roofing contractors, 230
Roofs, 250, 251
Rotary spreader, 109–11, *109, 111*
Rotating crops, 16
Rotenone, 63
Roundworms, 58–9, 61
Rove beetle, *69*
Ryania, 64
Ryegrass, *6*
  desired height, *8*

Sabadilla, 63–4
Safe Drinking Water Hotline, 180, 291
Safety
  consumer, 291
  disposal of auto products, 123–5
  disposal of paints and solvents, 128–30
  disposal of pesticides, 121–2
  fertilizers, 108–13
  hazardous materials, 116–20, *117–8*
  household cleaners, 247
  with pesticides, 97–107, *98–103*
  treated wood, 32
  See also Disposal; *specific topics*
Salt, 189–90, *192,* 248
Sanitation, 242–3
Sawdust, as mulch, 165
Screen cones, 78
Sealed combustion systems, 256
Sealing cracks, 214, *215*
Sealing homes, 258–65, *260–4*
Seattle, Washington, 231–3, *232–3*
Sedimentation zone, 184
Seeds, 9, 15, 35–6
Septic systems, 196–7, *196*

Septic tank, 196, *196*
Shower heads, 284, *284*
Shrubs
  native varieties, 21–6, *23–6*
  selecting, 21–6, *23–6*
Shutoff valves, 284, *284*
Silica aerogel, 244
Sill sealer, 251
Silverfish, *236*
Soil
  alkalinity, 21, 83
  building, 83
  compaction, 30
  core aerification, 7–8
  drainage, 17, 21
  earthworms, 8
  for gardens, 16–7
  pH, 17, 81, 163
  preparation for lawns, 4–5
  solarizing, 79–80
Solar heat. *See* Sun
Solarizing soil, 79–80
Solvents, 128–30, 132
Sowbug, *238*
Space heaters, 222–3, 225
Space sprays, 241
Spider, *238*
Spined soldier bug, *69*
Spray taps, 284, *284*
Sprayers, 101–7, *101–3*
Spreaders, 108–9, *108–11*
Squash, 20
Squirrels, *44,* 49
Staking, 37
Sticky traps, 78
Straw, as mulch, 165
Sub-slab suction, 212, *213,* 215
Sun, 279–81, *280*
Syrphid fly, *69*

Tall fescue, *6*
  desired height, *8*
Tar-paper disks, *77,* 78
Tarps, 79
Temperature control, 243
Termite, 54, *239*
Thatch, *7,* 7–8
  backyard prairies, 37
  core aerification, 7–8
  management, 154–5
  mechanical dethatching, 8
  power raking, 8
  vertical mowing, 8
Tires, 124–5
Toilets, *285,* 285–6
Toxic materials, 118

Toxicity, 92–4, *93, 98*
Traps
   for insects, 73–6, *74,* 245
   for nuisance animals, 47–50
Treated wood, 32
Tree residues, 166–7
Tree squirrels, 49–50
Trees, 293
   birds, 41
   construction damage, 30–1
   fertilization of, 27–8, *28*
   life spans, 27
   native varieties, 21–6, *22–3*
   pruning, 28
   selecting, 21–6, *22–3*
   for shade, 251, 279
   warning signs for stress, 28–30, *29*
   watering, 27
Trichogramma wasp, *69*
Trombone sprayers, 102
Turning bins, 160–1, *160–1*

U. S. Consumer Product Safety Commission, 291
U. S. Environmental Protection Agency
   address, 290
Ultrasonic pest repellers, *76*
Ultraviolet radiation, *192,* 193
Urban wildlife, 47
Urea formaldehyde, 227

Vacuum cleaners, 218
Varnish, 32
Vegetables
   disease-resistant, 20
   insect-resistant, 20
   pest control, 19–20, *19*
   See also Gardens
Ventilation, 226, 251, 253–4, 256, 265
Vinegar, 248
Viral insecticides, 57–8, *60*
Viruses
   insecticides, 57–8, *60*
VOCs, 130, 246
Volatile organic compounds, 130, 246

Walls, 256, 279–82, *280*
Waste
   generated in United States, *141*
   See also Recycling
Waste Watch Center, 293
Water, drinking, 291, 298–9
   boiling water, 175, 177
   bottled water, 198–200
   lead, 205–6

Safe Drinking Water Hotline, 180
   signs of contaminants, 186–7, *186–7*
   testing, 170–85, *175,* 198–9, 301
   types of, 201
   See also Water treatment
Water conservation, 283–8, *284–5,* 299–300
   indoors, 283–6, *284–5*
   outdoors, 286–7
Water samples, 173
Water-saving devices, 284, *284*
Water softening, 190, *192*
Water supply
   private versus public, 170–1
Water treatment, 182–5, *183,* 188–95, *189, 191–2*
   scams, 194–5
Water use, 283
Watering
   gardens, 18
   lawns, 7, 9, 286–7
   trees, 27
Waterproofing, 255
Weatherstripping, 258–62, *260–1*
   doors, *260*
   windows, *261*
Weeding, 37
Wells, 171, 174–6, *175,* 178–9, 188, 198, 201
Wild grasses, 36, *38–40*
   See also Native plants
Wildflowers, 36, *38–40*
   See also Native plants
Wildlife, urban, 47
Windbreaks, 251, 279, 281–2, *281*
Windows, 250, 256
   caulking, *262*
   glazing, 263–5, *263–4*
   heat loss, 263, 282
   insulation, 263, *263–4*
   U-values, *263,* 263
   weatherstripping, *261*
Wire-mesh bin, 160, *160*
Wood, 32, 166–7
Wood chips, as mulch, 163
Wood Heater Program, 290
Wood roach, *238*
Wood stoves, 222
Woodchucks, 50

Yard
   backyard prairie, 33–40, *38–40*
   waste, 143, 154–5, 162–3
   See also Lawn
Yellowjacket, *239*

Zeolite filter, 193
Zoysiagrass, 6
   desired height, *8*

This LAND

Beyond the home

This Land books provide ideas on how to protect your home environment, but they also reach the farm.

The first book in the series, *50 Ways Farmers Can Protect Their Groundwater,* was selected as top prize winner in an international agricultural publications awards program. Here's what others have to say about the book:

> *50 Ways...*"offers a thorough, yet practical checklist of voluntary practices designed to reduce the risk of groundwater contamination without cutting into yields or profitability." The recommendations in the book "not only save the environment but could possibly put dollars in your pocket."
> *Prairie Farmer* editors "found it both easy to read and well targeted.... The illustrations alone are worth the price of the book."
> —*Prairie Farmer*

> "Many of the practices described in *50 Ways* can boost profits by helping farmers cut back on chemical inputs....Sprinkled in among the how-to sections are profiles of farmers and their down-to-earth experiences integrating the practices into their farm management schemes."
> —Environmental Protection Agency's *News-Notes*

> *50 Ways...*"explains how people can live off their land without endangering what's below."
> —Associated Press Wire Service

> "From herbicides to livestock disposal, the book offers plain-spoken advice on issues that are certain to dominate agriculture the rest of the decade."
> —*The Milwaukee Sentinel*

To order your copy of *50 Ways Farmers Can Protect Their Groundwater,* or for additional copies of *57 Ways to Protect Your Home Environment (and Yourself),* contact:

University of Illinois
Information Services
69-BK Mumford Hall
1301 West Gregory Drive
Urbana, IL 61801
(217)333-2007